AF531001

High Altitude and Man

Christopher Pizzo, MD, taking alveolar gas samples on
the summit of Mount Everest, October 24, 1981, during the course of the
American Medical Research Expedition to Everest.

High Altitude and Man

EDITED BY John B. West

Section of Physiology
Department of Medicine
University of California, San Diego
La Jolla, California

Sukhamay Lahiri

Department of Physiology
Institute for Environmental Medicine
University of Pennsylvania School of Medicine
Philadelphia, Pennsylvania

AMERICAN PHYSIOLOGICAL SOCIETY
Bethesda, Maryland, 1984

Library of Congress Catalog Card Number 84-2933

International Standard Book Number 0-683-08945-5

Printed in the United States of America by
Waverly Press, Inc., Baltimore, Maryland 21202

Distributed for the American Physiological Society by
The Williams & Wilkins Company, Baltimore, Maryland 21202

Preface

There is a rich history on the subject of human physiology at high altitude. The wide-ranging features of acclimatization have long fascinated physiologists and physicians who see the topic as one of the best examples of how man can respond to a hostile environment. The subject has its own intrinsic scientific importance, and no one needs to justify an interest in it.

Nevertheless the last few years have seen a burgeoning of interest in the physiology of man at high altitude. There are several reasons for this. One of the most dramatic was the ascent of Mount Everest by two climbers without supplementary oxygen in 1978, a feat that many physiologists thought was impossible. This provoked a great deal of interest in the physiology of extreme altitudes. Another reason is the increasing concern in how to improve man's well-being and physical performance at high altitude. A large number of people in the world are natives to altitudes over 3,000 m. Moreover in recent years a substantial number of people have moved to these altitudes, and with increasing industrialization in those areas, man's ability to carry out physical work becomes a matter of economic importance. In addition the armed forces are concerned about how to maximize human performance under these conditions. Finally, the study of man at high altitude provides unique information about the effects of severe hypoxia, which are clearly relevant to the pathophysiology of patients with lung and heart disease.

This monograph is an outgrowth of a symposium on man at high altitude sponsored by the American Physiological Society. The symposium was in three parts, covering the topics of man at extreme altitude, sleep and respiration at high altitude, and physiology of permanent residents of high altitude. The timing of the symposium was stimulated in part by the American Medical Research Expedition to Everest, which took place in the fall of 1981. Although the symposium was not solely devoted to results obtained by the expedition, a number of papers stemmed from it, and we have included a brief introduction to give some background to the expedition.

The Editors

Contents

Physiology of Permanent Residents of High Altitude

Introduction: American Medical Research Expedition to Everest

JOHN B. WEST

Section of Physiology, Department of Medicine, University of California, San Diego, La Jolla, California

SEVERAL CHAPTERS OF THIS BOOK describe work carried out during the course of the 1981 American Medical Research Expedition to Everest. The purpose of this brief introduction is to give some of the background to this unusual expedition.

Expeditions to high altitude have traditionally contributed a great deal to our understanding of high-altitude physiology. Some of the most memorable examples include the International Expedition to Pike's Peak in 1911, with its controversial evidence for oxygen secretion; Barcroft's expedition to Cerro de Pasco, Peru, in 1921–1922, which first drew attention to the remarkable adaptation of high-altitude natives; the International High Altitude Expedition to Aucanquilcha in the Chilean Andes in 1935, which laid the foundations of our knowledge of blood biochemistry at high altitude; and the Himalayan Scientific and Mountaineering Expedition of 1960–1961, which obtained the first measurements in man above 6,000 m.

The principal objective of the 1981 American Medical Research Expedition to Everest was to obtain information on human physiology at extreme altitudes, including the first measurements over 8,000 m. In line with this objective, the design of the expedition was very unusual. First, there was a group of six experienced Himalayan climbers, including John P. Evans, Climbing Leader. Next, there were six "climbing scientists," all of whom were strong climbers, but each was a medical doctor with an interest in high-altitude physiology. Their responsibility was to obtain the data at extreme altitudes. Finally, there was a third group of eight physiologists who worked in the two laboratories at Camp II (6,300 m) and Base Camp (5,400 m). The expedition was supported by 42 high-altitude Sherpas.

Figure 1 shows the four sites on the southern approach to Mount Everest at which physiological measurements were made. The Base Camp is readily accessible via the Khumbu Glacier. The laboratory there was a rigid prefabricated hut (Fig. 2) that was the site of an extensive research program during the months of September and October. Above the Base Camp is the steep and treacherous Khumbu Icefall, which leads into a high, relatively level valley, the Western Cwm. This was the site of the Main Laboratory Camp, altitude 6,300 m (Fig. 3). This laboratory was constructed of fiberglass blankets covering an aluminum frame, and there was a plywood floor. Both laboratories

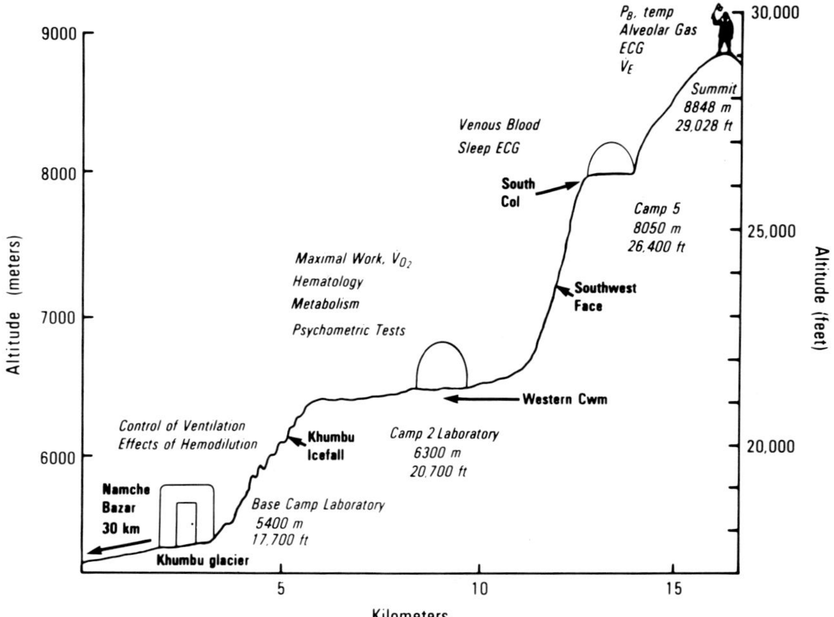

FIG. 1. Research program of the 1981 American Medical Research Expedition to Everest, indicating the four sites on the mountain at which measurements were made.

FIG. 2. Prefabricated hut used as the physiological laboratory at Base Camp (5,300 m).

FIG. 3. Main Laboratory at Camp II, 6,300 m. Structure has a plywood floor and an aluminum frame that is covered by fiberglass blankets.

were kept warm with propane heaters and provided with electrical power from solar panels and gasoline generators.

Beyond the Western Cwm is a steep headwall, which leads to a flat area, the South Col. The highest camp, Camp V, was pitched a little above this, at an altitude of 8,050 m. Here the climbers were instrumented before going to the summit, samples of venous blood were taken, and sleep electrocardiograms were measured.

The most ambitious part of the scientific program was the measurements scheduled for the summit. Here we were extremely fortunate; three Americans and two Sherpas reached the summit. Two of the Americans, Christopher Pizzo, MD, and Peter Hackett, MD, were "climbing scientists" and made measurements en route. Pizzo succeeded in measuring barometric pressure and temperature on the summit and in taking a series of alveolar gas samples (see frontispiece). This was a magnificent achievement.

In summary, the expedition was an outstanding success. An extensive scientific program was completed. Christopher Kopczynski, Christopher Pizzo, and Peter Hackett were the ninth, tenth, and eleventh Americans to reach the summit. Their Sherpa companions were Sungdare and Young Tensing. Finally, everyone returned home safely. We hope that this book is a useful record of some of the physiological research (though by no means all) carried out during the expedition.

The Expedition was sponsored by the American Alpine Club, the American Physiological Society, and the Explorers Club. The last also made a contribution toward the costs of publication of this monograph.

1

Man on the Summit of Mount Everest

JOHN B. WEST

Section of Physiology, Department of Medicine, University of California, San Diego, La Jolla, California

Barometric Pressures on Mount Everest
Alveolar Gas Composition
Partial Pressure of Arterial Oxygen
Acid-Base Status
Maximal Exercise

PHYSIOLOGISTS AND PHYSICIANS have long been interested in the physiology of man at extreme altitude because of the problems that this environment poses for adequate tissue oxygenation. Indeed it is arguable that exercise at great altitude is the most demanding respiratory challenge that normal man can encounter. There have been a number of instances where ascents to high altitude have prompted changes in thinking about severe hypoxia. For example, when the Duke of the Abruzzi reached the extraordinary altitude of 7,500 m in the Karakoram Mountains in 1909 (9), Douglas and his colleagues (7) concluded that adequate oxygenation of the blood would be impossible under these conditions without active secretion of oxygen. This notion was reinforced by measurements made during the International Pike's Peak Expedition of 1911 (7) but laid to rest in part by Barcroft and his colleagues (1) during their expedition to Cerro de Pasco in Peru in 1921–1922 and the International High Altitude Expedition to the Chilean Andes in 1935 (6).

More recently the ascent of Mount Everest (8,848 m) without supplemental oxygen by Messner (15) and Habeler (12) in 1978 stimulated renewed interest in the effects of the very severe hypoxia of extreme altitudes on normal man. As an example, Figure 1 shows maximum oxygen uptake at various barometric pressures in acclimatized subjects measured by Pugh et al. (20). Extrapolation of the line to the barometric pressure at the summit of Mount Everest (~250 mmHg) suggests that nearly all the available oxygen is required for basal metabolism (31).Thus it appears that man breathing ambient air at these extreme altitudes is critically near his limit. The evolutionary pressures responsible for man being just able to reach the highest point on earth are certainly obscure!

One of the chief objectives of the 1981 American Medical Research Expedition to Everest (27) was to obtain additional data on the physiology of man at extreme altitudes. We had previously carried out a theoretical study (31) in an effort to identify the most important variables. It was clear from

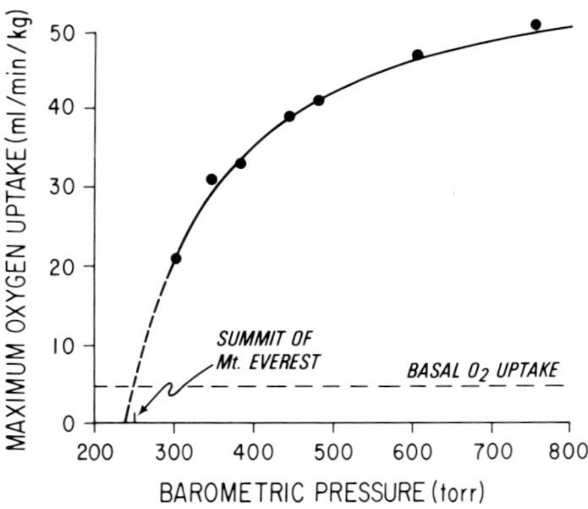

FIG. 1. Maximum O_2 consumption in acclimatized subjects plotted against barometric pressure. Predicted maximum uptake on the summit of Mount Everest is very close to basal O_2 requirements. [Adapted from West and Wagner (31); data from Pugh et al. (20).]

this that maximal oxygen uptake under these remarkable conditions would be exquisitely sensitive to barometric pressure and to the level of hyperventilation. One surprising prediction was that respiratory alkalosis would apparently enhance oxygen transfer to the tissues, as judged by the partial pressure of oxygen (PO_2) of mixed venous blood. The reason was that the left-shifted oxygen-dissociation curve would improve loading of oxygen in the pulmonary capillary (2). Dejours (5) has independently done another theoretical study with somewhat different results.

BAROMETRIC PRESSURES ON MOUNT EVEREST

Because barometric pressure was clearly critical, we took great pains to measure it accurately at several altitudes, including the summit (30). At Base Camp, altitude 5,400 m, the measurements were made with a Fortin mercury barometer especially shortened for the expedition so that the low pressures fell on scale. The barometer was kept in a warm laboratory and read each morning. Above Base Camp most of the measurements were made with a small portable barometer also especially designed for the expedition. The pressure transducer was a crystal with one side exposed to a vacuum and the other to the air pressure; it had a digital output and a high intrinsic accuracy and was checked against the mercury barometer kept at Base Camp. A few measurements above Base Camp were also made with a highly accurate aneroid barometer.

Figure 2 shows some of the results. At Base Camp (5,400 m) the mean pressure was 400.4 mmHg (n = 35; SD = 2.7); this fell to 283.6 mmHg (n = 6; SD = 1.5) at Camp V (8,050 m). The one measurement obtained on the summit was 253.0 mmHg. The altitudes of these three sites are accurately known from surveys. All these pressures are substantially above those predicted from the International Civil Aviation Organization (ICAO) Standard Atmosphere (13) for these altitudes (Fig. 2). For example, the ICAO Standard Atmosphere pressure for an altitude of 8,848 m is only 236 mmHg, which is 17 mmHg less than our reading. The Standard Atmosphere (ICAO and US are essentially identical in this range) has been used extensively by physiologists and physicians to predict the deleterious effects of hypoxia at great altitudes, and it is universally employed to calibrate high-altitude chambers. These disparities are therefore important. That barometric pressures on mountains often exceed those predicted from the Standard Atmosphere has been previously recognized (18).

Figure 3 shows why barometric pressures at these altitudes are consider-

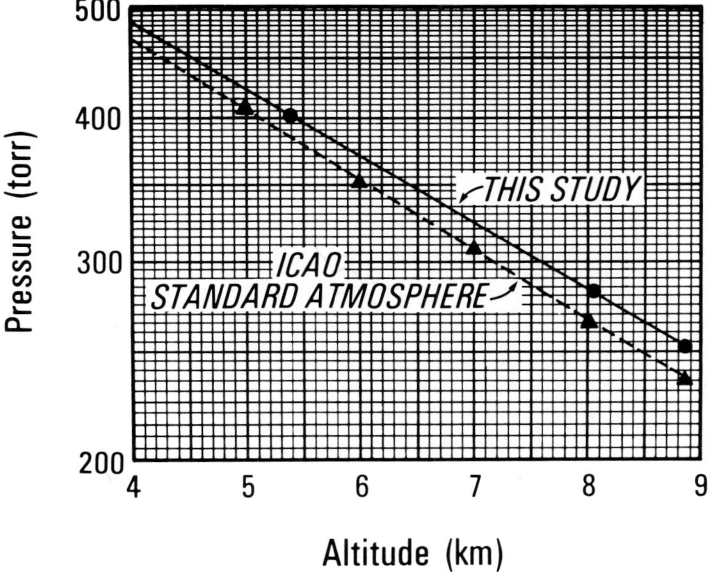

FIG. 2. Barometric pressure as found on expedition plotted against altitude. Pressures are substantially higher than those predicted by International Civil Aviation Organization (ICAO) Standard Atmosphere (13). [From West et al. (30).]

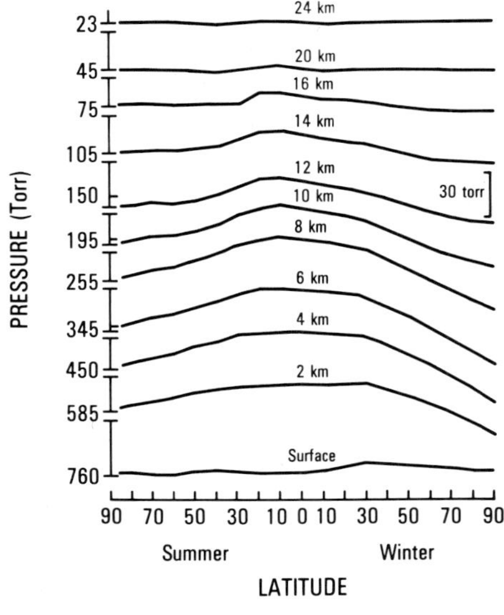

FIG. 3. Increases of barometric pressure near equator at various altitudes (summer and winter). *Vertical scale*, pressure increasing upward according to *scale* on *right*. *Numbers* on *left*, barometric pressures at poles for various altitudes. [Adapted from Brunt (3).]

ably higher than the Standard Atmosphere predictions. Although barometric pressures at the surface of the earth and at an altitude of 24 km are essentially independent of latitude, pressures in the range of 4–16 km are markedly latitude dependent. This is because of a large mass of very cold air in the stratosphere above the equator (paradoxically the coldest air in the atmosphere) that is caused by complex radiation and convection phenomena (3). Theoretical studies clearly show that if the barometric pressure at the summit of Mount Everest were not increased by this equatorial bulge, there would be barely enough oxygen available to sustain human basal metabolism (31). Thus this climatic idiosyncrasy makes it possible for man to reach the highest point on earth while breathing ambient air.

ALVEOLAR GAS COMPOSITION

To obtain information about pulmonary gas exchange at these extreme altitudes, we designed an automatic alveolar gas sampler (see frontispiece). The subject exhaled to residual volume through the mouthpiece and held his breath for a second or so. When he pulled the lever, a valve of a small preevacuated aluminum can was opened, thus trapping the last expired gas. When the lever was released, the can valve was closed and a cartridge revolved to bring another sampling can into the correct position. In this way the subject could take six alveolar gas samples without complicated manipulations (14).

Thirty-four valid alveolar gas samples were collected above an altitude of 8,000 m (29). The mean P_{CO_2} at Camp V (altitude 8,050 m) was 11.0 ($n = 27$; SD = 2.0); this had fallen to a mean value of 8.0 ($n = 3$) at an altitude of 8,400 m. The mean of four valid samples on the summit was 7.5 mmHg. These astonishingly low values indicate the extreme hyperventilation characteristic of successful climbers at these altitudes.

Figure 4 shows the mean P_{CO_2} values (▲) and other data from previous

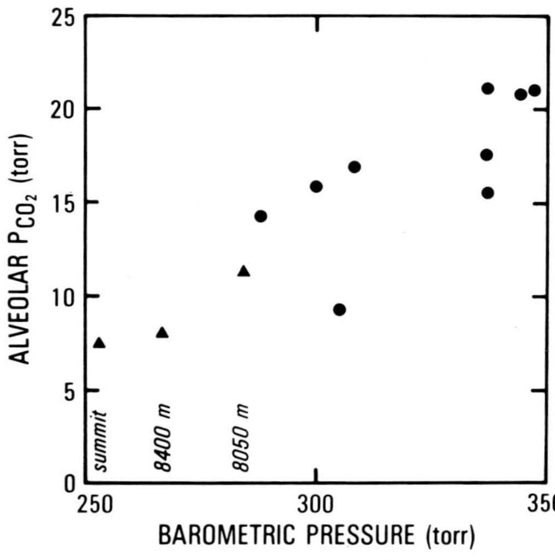

FIG. 4. Partial pressure of CO_2 (P_{CO_2}) in alveolar gas plotted against barometric pressure. ▲, Means of measurements made on expedition; ●, results of previous investigators at barometric pressures of <350 mmHg. [From West et al. (29).]

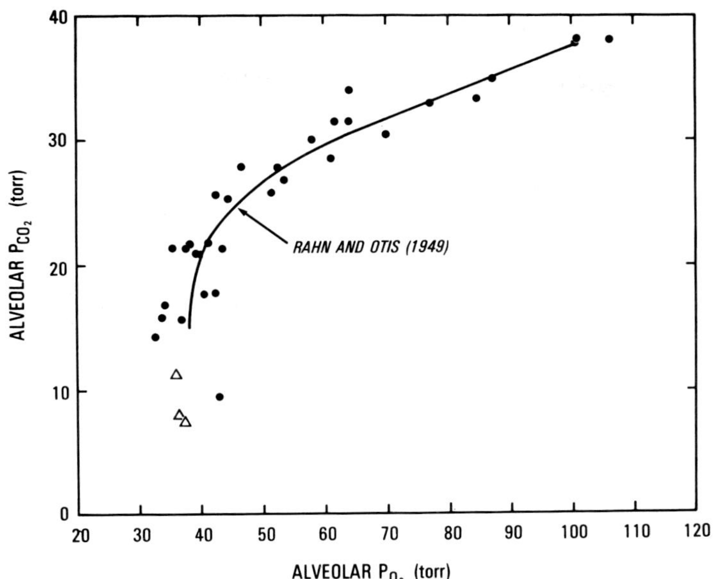

FIG. 5. O_2-CO_2 diagram showing alveolar gas composition in acclimatized subjects at high altitude. △, Means of measurements made on expedition; ●, results of previous investigators. At extreme altitudes the marked hyperventilation maintains alveolar P_{O_2} at approximately 35 mmHg. [From West et al. (29).]

expeditions at barometric pressures below 350 mmHg (●). Note the approximately linear decrease in alveolar P_{CO_2} with decreased barometric pressure and that our results appear to be consistent with previous measurements. However, our summit samples were marred by high respiratory exchange ratio (R) values (mean 1.48). The reasons for this are not clear, but the values indicate that Pizzo, who made the measurements, was not in a steady state. He reported that the effort of delivering the samples was considerable under those conditions, and remarks recorded by him on a small dictating tape recorder on the summit clearly confirm that he was very short of breath. Also he had an extensive program to complete in a short time. However, the samples taken at an altitude of 8,400 m were satisfactory, with a mean R value of 0.75. Pizzo had much more time to take these because he had to wait several hours for another climber, Hackett, to return from a summit bid. Because the mean P_{CO_2} at that altitude was only 8.0 mmHg, we can be confident that the P_{CO_2} on the summit was lower.

Also note that Pizzo used supplementary oxygen en route to the summit, which increased his inspired P_{O_2} to approximately 70 mmHg. The oxygen mask was removed for at least 10 min before the alveolar gas samples were taken, but the results may have been affected. If oxygen had any influence, it presumably depressed ventilation to some extent. Therefore we can conclude that the alveolar P_{CO_2} of a climber on the summit who did not use oxygen would not be higher than 7.5 mmHg but might be lower.

Figure 5 shows the results of the alveolar gas analyses plotted on an O_2-

CO_2 diagram. Again the means of the samples were obtained on our expedition (△) and the data were obtained by previous investigators on subjects acclimatized to high altitude (●). The data are from various sources (10, 11, 18, 25) and were graphed in this way by Rahn and Otis (21). The line of best fit that they drew is indicated. Note that our data points are generally consistent with an extrapolation of this line.

An interesting feature of Figure 5 is that an imaginary line of best fit through the data has an almost vertical slope at the highest altitudes (lowest values of P_{CO_2}). This means that the body responds to lower and lower barometric pressures by increasing ventilation to such an extent that the alveolar P_{O_2} is successfully defended at a value of 35 mmHg. Thus, although the P_{CO_2} continues to fall as altitude is increased, the alveolar P_{O_2} shows an almost constant value. Incidentally, probably not all subjects can respond to increasing altitudes with this degree of hyperventilation. However, not everyone can reach the summit of Mount Everest, and these two facts are probably connected.

PARTIAL PRESSURE OF ARTERIAL OXYGEN

Naturally it would be interesting to have information on the arterial P_{O_2} of a climber on the summit. Unfortunately it is impractical to take arterial blood under these conditions. However, we can get some information by calculating the increase in P_{O_2} along the pulmonary capillary with the available data. The procedure closely follows that used to predict the gas exchange on the Everest summit (31). Table 1 shows the measured and assumed values. The hemoglobin concentration, P_{50}, and base excess were determined from venous blood samples taken from the two medical summitters, Pizzo and Hackett, during the morning after their successful summit climb. As indicated below, there is good evidence that base excess was changing very slowly at these altitudes. The assumptions for the Bohr integration procedure include reasonable values for oxygen uptake (350 ml/min) and respiratory exchange

TABLE 1. *Values used for calculating partial pressure of oxygen in arterial blood on summit of Mount Everest*

Measured	
Barometric pressure	253 mmHg
Partial pressure of CO_2 in alveolar gas	7.5 mmHg
Hemoglobin concentration	18.4 g/dl
P_{50} at pH 7.4	29.6 mmHg
Base excess	−7.2 meq/liter
Assumed	
Oxygen uptake	350 ml/min
Respiratory exchange ratio	0.85
Cardiac output	6 liters/min
Membrane diffusing capacity for O_2	40 $ml \cdot min^{-1} \cdot mmHg^{-1}$
Pulmonary capillary blood volume	75 ml

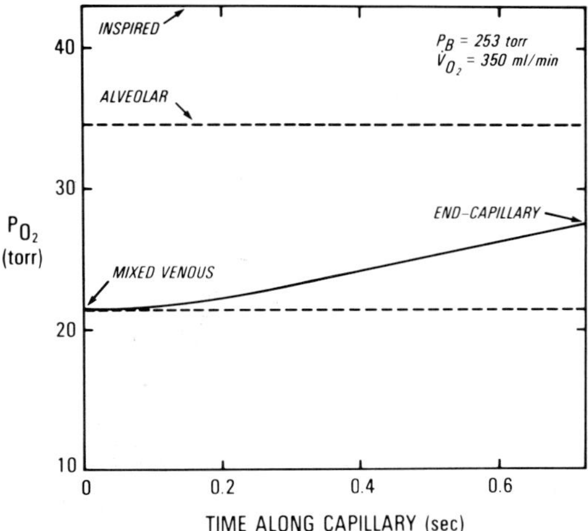

FIG. 6. Calculated changes in P_{O_2} along pulmonary capillary for a climber at rest on the summit of Mount Everest. P_{O_2} in mixed venous blood is 21 mmHg; this rises slowly along the capillary to reach only 28 mmHg at the end. Large P_{O_2} difference of 7 mmHg between alveolar gas and end-capillary blood indicates marked diffusion limitation of O_2 transfer. $\dot{V}_{O_2}$, O_2 consumption; P_B, barometric pressure. [From West et al. (29).]

ratio (0.85). A normal resting cardiac output of 6 liters/m was assumed from measurements by Pugh (19) in acclimatized subjects at an altitude of 5,800 m. He showed that the relationship between cardiac output and oxygen uptake under these conditions was the same as at sea level; Cerretelli (4) reported results consistent with this. Finally, a normal membrane diffusing capacity for oxygen and pulmonary capillary blood volume were assumed from measurements of carbon monoxide diffusing capacity made in acclimatized subjects at an altitude of 5,800 m (26).

Figure 6 shows the results of the calculation for a climber on the summit. The alveolar P_{O_2} was approximately 35 mmHg, being set by the inspired P_{O_2}, alveolar P_{CO_2}, and R values. To satisfy the oxygen uptake of 350 ml/min, the P_{O_2} in mixed venous blood was approximately 21 mmHg. A striking feature of the analysis was that the P_{O_2} rose very slowly along the pulmonary capillary. The P_{O_2} in end-capillary and therefore arterial blood was approximately 28 mmHg. Thus the P_{O_2} difference between alveolar gas and end-capillary blood was about 7 mmHg. Clearly oxygen transfer was markedly diffusion limited under these conditions. The importance of this diffusion limitation of oxygen transfer has been emphasized recently in other studies (17, 24, 31).

The assumed resting value for membrane diffusing capacity for oxygen of 40 $ml \cdot min^{-1} \cdot mmHg^{-1}$ (corresponding to a pulmonary diffusing capacity of 33.6 $ml \cdot min^{-1} \cdot mmHg^{-1}$) agrees with several studies. However, because some argue that it may be higher (16), additional calculations were made for membrane diffusing capacities of 60 and 80 $ml \cdot min^{-1} \cdot mmHg^{-1}$. These increased the calculated P_{O_2} of end-capillary blood to 31 and 32 mmHg, respectively.

ACID-BASE STATUS

What can be said about the acid-base status of the arterial blood of a climber on the summit? The venous blood samples taken from Pizzo and

Hackett the morning after the successful summit climbs were put on ice in insulated Dewar flasks and taken down to the Main Laboratory (6,300 m) within a few hours. The samples were tonometered against a known P_{CO_2} (22.6 mmHg), and a mean pH of 7.442 was measured (32).

Now if we assume that there was no change in base excess during the 19–22 h that elapsed between when the subjects were on the summit and when the blood samples were taken, the Siggaard-Andersen nomogram (23) can be used to calculate the arterial pH on the summit. The arterial P_{CO_2} is assumed to be the same as the alveolar value, that is, 7.5 mmHg. Figure 7 shows the results of this procedure for Pizzo, whose P_{CO_2} was measured on the summit.

The lowest line joins the tonometer values of 22.6 mmHg for P_{CO_2} and 7.46 for pH. By using the pivot point corresponding to Pizzo's hemoglobin of 18.7 g/dl and his alveolar P_{CO_2} of 12.5 mmHg at 8,050 m, a second line is drawn. This corresponds to the in vitro blood buffer line because the blood passing through the pulmonary capillary does not have access to the extravascular interstitial buffering pool (22). Finally, a third line is drawn from the summit P_{CO_2} of 7.5 mmHg to the pivot point corresponding to one-third of the hemoglobin concentration, this being the in vivo buffer line (22). The resultant arterial pH is about 7.76. No allowance has been made for the effect on base excess of differences in the oxygen saturation of the blood on the summit and in the tonometer because these were almost the same (32). Note

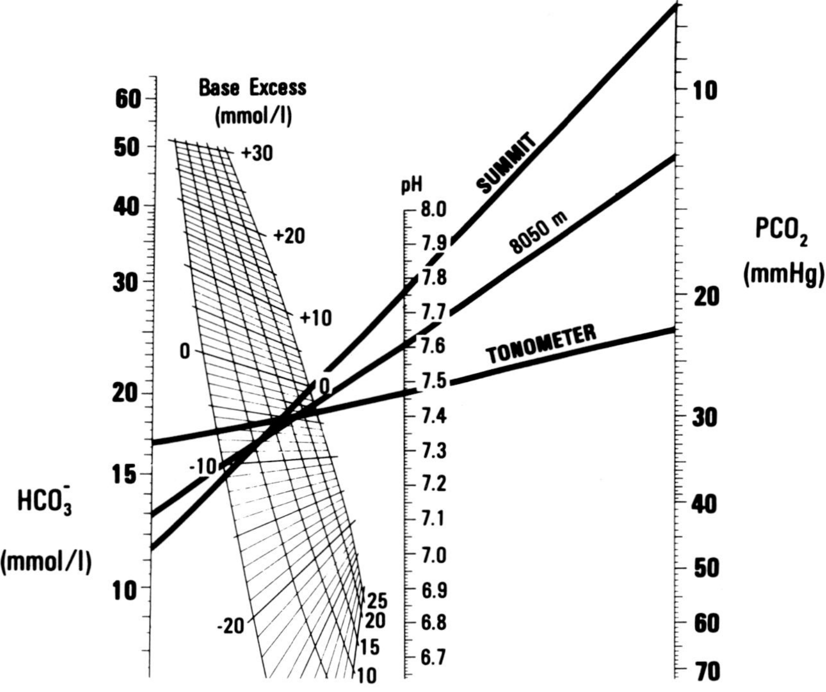

FIG. 7. Siggaard-Andersen nomogram (22) showing results obtained from tonometered venous blood of Pizzo. For a P_{CO_2} of 7.5 mmHg on the summit, calculated arterial pH is about 7.76 (see text for details).

that the nomogram (or equations on which it is based) cannot be expected to give accurate results at these extreme values.

As indicated above, an assumption of this calculation is that there was no change in base excess between when the climbers were on the summit and when the blood samples were taken. The available data suggest that there was little, if any, change. For example, measurements carried out on other members of the expedition (32) showed paradoxically that Pizzo and Hackett had smaller reductions in base excess when the measurements were taken at Camp V (8,050 m) than most of the expedition members who were living at Camp II (6,300 m) (32). Thus the average base excess of the subjects at Camp II was −8.7 meq/liter, whereas the values for Pizzo and Hackett were −5.9 and −8.4 meq/liter, respectively. We also have data on Pizzo that show (again paradoxically) that when he was at Camp II, 15 days prior to his successful summit climb, his base excess was −9.8 meq/liter, whereas, as indicated here, his base excess on the morning after the summit ascent was 5.9 meq/liter. Thus these data certainly do not suggest that Pizzo reduced his base excess any more as a result of his excursion to the summit.

The calculation of arterial pH in a climber resting on the summit neglects the effects of any residual lactic acid in the blood as a result of the summit climb. We have no way of allowing for this. However, note that the increase in blood lactate after maximal exercise at high altitude is much lower than at sea level (8). In fact measurements made on three subjects at an altitude of 6,300 m showed that maximal exercise increased blood lactate by a mean value of only 11.7 mg/dl (28). It should also be remembered that the climbers were breathing supplemental oxygen during their summit climb. Naturally, insofar as lactate was present in the blood, this would reduce the high calculated pH.

The extreme respiratory alkalosis resulted in a calculated arterial oxygen saturation of over 70% on the summit despite the low arterial PO_2 of 28 mmHg. Surprisingly the oxygen saturation on the summit was nearly the same as the average arterial oxygen saturation of 72% measured on arterial samples at Camp II (6,300 m) (32). Table 2 summarizes the summit values and compares them with normal values at sea level.

MAXIMAL EXERCISE

So far this analysis has been confined to resting conditions at extreme altitude. What are the consequences of increasing oxygen uptake by exercise

TABLE 2. *Alveolar gas and arterial blood values for subject CP on summit of Mount Everest*

Altitude	Barometric Pressure, mmHg	Partial Pressure, mmHg				Arterial pH
		Inspired O_2	Alveolar O_2	Arterial O_2	Arterial CO_2	
Summit, 8,848 m	253	43	35	28	7.5	7.76
Sea level	760	149	100	95	40	7.40

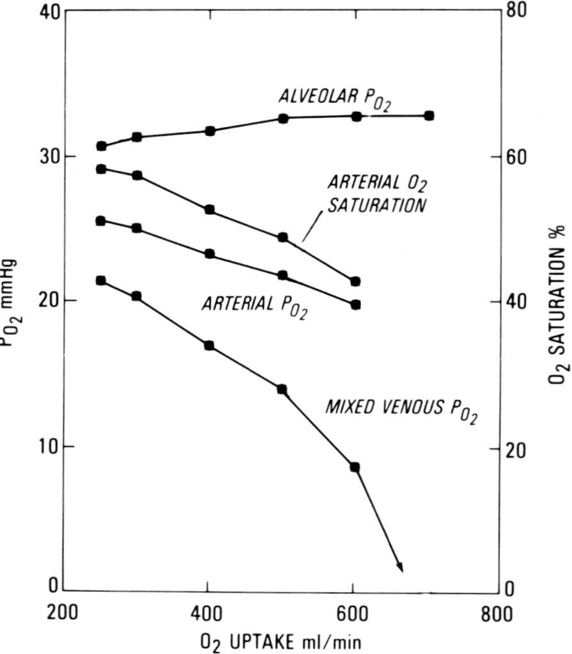

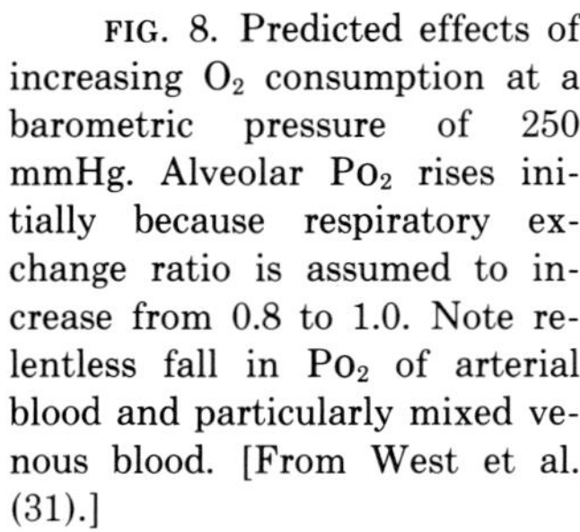
FIG. 8. Predicted effects of increasing O_2 consumption at a barometric pressure of 250 mmHg. Alveolar Po_2 rises initially because respiratory exchange ratio is assumed to increase from 0.8 to 1.0. Note relentless fall in Po_2 of arterial blood and particularly mixed venous blood. [From West et al. (31).]

under these conditions? As shown in Figure 6, there is evidence of marked diffusion limitation of oxygen transfer across the pulmonary capillary at rest. A characteristic of the hypoxemia caused by diffusion limitation is that it is much exaggerated during exercise. Figure 8 shows the predicted results of increasing oxygen consumption at a barometric pressure of 250 mmHg (31). Note the relentless fall in arterial Po_2 with increasing oxygen uptake and the widening alveolar-arterial Po_2 difference. The slight increase in alveolar Po_2 is the result of a small increase in the R value from 0.8 to 1. Even more impressive than the fall in arterial Po_2 is the steep reduction in the Po_2 of mixed venous blood. If we take the mixed venous Po_2 as generally indicative of tissue Po_2 and assume that there is a value below which further increases of oxygen uptake are not possible, Figure 8 clearly implies that maximal exercise is limited by oxygen diffusion across the blood-gas barrier. Indeed, if we take the minimal Po_2 of mixed venous blood to be 15 mmHg, the conditions of Figure 8 predict a maximal oxygen uptake ($\dot{V}o_{2\,max}$) of less than 500 ml/min. If the membrane diffusing capacity for oxygen is increased to 100 ml$\cdot$min$^{-1}\cdot$mmHg^{-1}, the value of $\dot{V}o_{2\,max}$ increases less than 750 ml/min (31).

An extensive series of measurements of maximal exercise was made during the expedition (28). We used a bicycle ergometer and generally required the subjects to maintain a particular work rate for 5 min, though at the higher work levels, especially at high altitude, this period was reduced to 3 min. The measurements were carried out in the Main Laboratory at Camp II (6,300 m), which was kept warm with a propane heater. Fourteen subjects were studied during ambient-air breathing when the inspired Po_2 was 63.7 mmHg. Eight subjects were able to sustain a work level of 1,200 $kg \cdot m^{-1} \cdot min^{-1}$. Six subjects were studied while breathing a 16% oxygen mixture, which gave an inspired Po_2 of 48.5 mmHg. Two subjects performed maximal exercise while breathing 14% oxygen, which gave an inspired Po_2 of 42.5 mmHg; this was equivalent to that on the summit of Mount Everest. Base-line measurements at sea level were obtained on all subjects prior to the expedition.

Arterial oxygen saturations were measured with a Hewlett-Packard ear oximeter. This was directly calibrated against arterial blood samples in the laboratory at 6,300 m. Figure 9 shows arterial oxygen saturation plotted against work rate at sea level prior to the expedition and on acclimatized subjects at an altitude of 6,300 m while they were breathing the three different gas mixtures. Note the progressive fall in arterial oxygen saturation with increasing work rate for the three levels of low inspired Po_2. This is in line with the predictions in Figure 8 and is consistent with marked diffusion limitation of oxygen transfer by the lung.

Figure 10 summarizes the oxygen uptake measured during maximal exercise measurements and compares our results with those found by Pugh and his co-workers (20) during the Himalayan Scientific and Mountaineering Expedition of 1960–1961. To our knowledge these are the only other measurements at such extreme altitudes. At the outset, it should be noted that our subjects had a higher maximal work capacity at sea level because many were unusually athletic. For example, four were competitive marathon runners and two were strong long-distance runners. Note that at an altitude of 6,300 m (barometric pressure of 351 mmHg; inspired Po_2 of 63.7 mmHg) $\dot{V}o_{2\,max}$ was reduced to 2.3 liters/min, which is about 50% of the sea-level value of 4.6 liters/min. When 14% oxygen was breathed at this altitude (inspired Po_2 of 42.5 mmHg) $\dot{V}o_{2\,max}$ was reduced to 1.1 liters/min, which is about 23% of the sea-level value. Figure 10 shows that the slope of the line relating $\dot{V}o_{2\,max}$ to inspired Po_2 at extreme altitudes was similar in both studies but that our points were shifted slightly to the left. Because of the steepness of the slope,

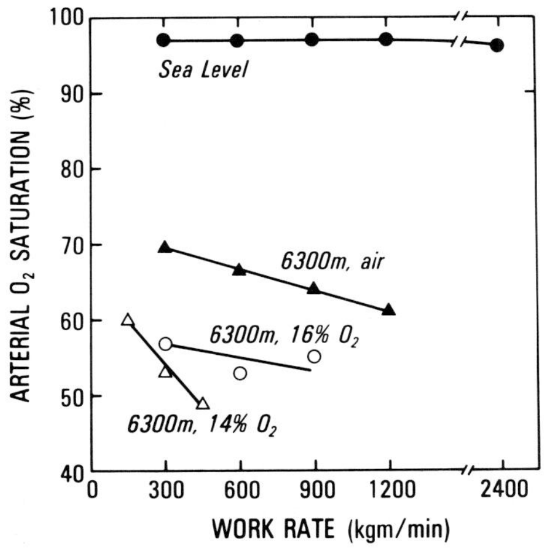

FIG. 9. Arterial O_2 saturation plotted against work rate for 4 values of inspired Po_2. Note steeply falling saturation as work rate increased when inspired Po_2 was low. This is consistent with diffusion limitation of O_2 transfer across blood-gas barrier (cf. Fig. 8). [From West et al. (28).]

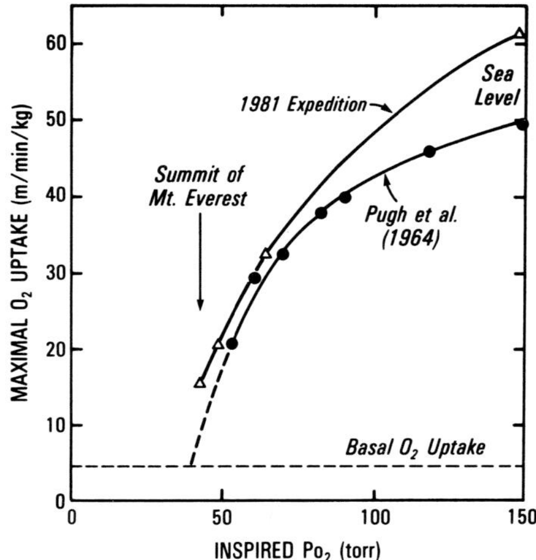

FIG. 10. Maximal O_2 uptake plotted against inspired P_{O_2} as measured during expedition (△). Previous data at extreme altitudes are from Pugh et al. (20) (cf. Fig. 1). Although *curve* derived from data of 1981 expedition is only slightly shifted to *left*, gain in maximal O_2 uptake at extreme altitudes is substantial. [From West et al. (28).]

this results in a substantial increase in oxygen uptake at these extreme altitudes.

Although a $\dot{V}O_{2\,max}$ of 1 liter/m is clearly limiting, it seems sufficient to explain how Messner and Habeler reached the summit of Mount Everest without supplementary oxygen. Messner (15) states that "the last 100 m of height took us more than an hour to climb." Assuming a climbing rate of 2 m/min and a total weight (climber and equipment) of 100 kg, the average work rate was 200 $kg \cdot m^{-1} \cdot min^{-1}$. This is appreciably less than the 300 $kg \cdot m^{-1} \cdot min^{-1}$ corresponding to the $\dot{V}O_{2\,max}$ of 1.1 liters/min measured on the two subjects inspiring a P_{O_2} of 42.5 mmHg. In fact both subjects were able to sustain a work rate of 450 $kg \cdot m^{-1} \cdot min^{-1}$ for 3 min although oxygen uptake did not increase, indicating that they were merely incurring a larger oxygen debt.

The improved performance of the subjects on the 1981 American Medical Research Expedition to Everest over that predicted from our previous theoretical analysis (31) can be explained by three main factors. *1*) The barometric pressure on the summit was slightly higher than we had predicted. *2*) The base excess of our subjects was substantially higher (less negative) than we had expected. Indeed the expedition members who spent 3–4 wk at Camp II (6,300 m) never regained their normal pH (i.e., respiratory alkalosis was never fully compensated), and there was apparently no further fall in base excess when the climbers went higher. The result was a markedly elevated pH at extreme altitude, and it can be shown that, when oxygen transfer in the lung is diffusion limited, the left-shifted oxygen-dissociation curve enhances the loading of oxygen in the pulmonary capillaries, resulting in a higher tissue P_{O_2} (2, 31). *3*) The level of hyperventilation was much higher than expected. This is the most important factor. We had previously assumed a P_{CO_2} of 10 mmHg on the summit, whereas the actual value was about 7.5 mmHg. The additional ventilation results in striking gains in arterial and therefore tissue P_{O_2}. It is a combination of these three factors that just allows man to reach the highest point on earth.

The expedition was supported by Public Health Service Grant R01-HL-24335 and Contract N01-HR-2915 and grants from the American Alpine Club, American Lung Association, National Geographic Society, National Science Foundation, Servier Laboratories (Paris), the Explorers Club, and the United States Army Research and Development Command.

REFERENCES

1. BARCROFT, J., C. A. BINGER, A. V. BOCK, J. H. DOGGART, H. S. FORBES, G. HARROP, J. C. MEAKINS, AND A. C. REDFIELD. Observations upon the effect of high altitude on the physiological processes of the human body, carried out in the Peruvian Andes, chiefly at Cerro de Pasco. *Phil. Trans. R. Soc. London Ser. B* 211: 351–480, 1923.
2. BENCOWITZ, H. Z., P. D. WAGNER, AND J. B. WEST. Effect of change in P_{50} on exercise tolerance at high altitude: a theoretical study. *J. Appl. Physiol.: Respirat. Environ. Exercise Physiol.* 53: 1487–1495, 1982.
3. BRUNT, D. *Physical and Dynamical Meteorology* (2nd ed.). Cambridge, UK: Cambridge Univ. Press, 1952, p. 379.
4. CERRETELLI, P. Limiting factors to oxygen transport on Mount Everest. *J. Appl. Physiol.* 40: 658–667, 1976.
5. DEJOURS, P. Mount Everest and beyond: breathing air. In: *A Companion to Animal Physiology*, edited by C. R. Taylor, K. Johansen, and L. Bolis. New York: Cambridge Univ. Press, 1982, p. 17–30.
6. DILL, D. B., E. R. CHRISTENSEN, AND H. T. EDWARDS. Gas equilibria in the lungs at high altitudes. *Am. J. Physiol.* 115: 530–538, 1936.
7. DOUGLAS, C. G., J. S. HALDANE, Y. HENDERSON, AND E. C. SCHNEIDER. Physiological observations made on Pike's Peak, Colorado, with special reference to adaptations to low barometric pressures. *Phil. Trans. R. Soc. London Ser. B* 203: 185–318, 1913.
8. EDWARDS, H. T. Lactic acid in rest and work at high altitude. *Am. J. Physiol.* 116: 367–375, 1936.
9. FILIPPI, F. D. *Karakoram and Western Himalaya 1909.* London: Constable, 1912.
10. GILL, M. B., J. S. MILLEDGE, L. G. C. E. PUGH, AND J. B. WEST. Alveolar gas composition at 21,000 to 25,700 ft. (6400–7830 m). *J. Physiol. London* 163: 373–377, 1962.
11. GREENE, R. Observations on the composition of alveolar air on Everest, 1933. *J. Physiol. London* 32: 481–485, 1934.
12. HABELER, P. *Everest: Impossible Victory.* London: Arlington Books, 1979.
13. *Manual of the ICAO Standard Atmosphere* (2nd ed.). Montreal, Canada: Int. Civ. Aviat. Org., 1968.
14. MARET, K. Expedition to Mt. Everest, 1981: technical aspects. In: *Hypoxia: Man at Altitude*, edited by J. R. Sutton, N. L. Jones, and C. S. Houston. New York: Thieme-Stratton, 1982.
15. MESSNER, R. The mountain. In: *Everest: Expedition to the Ultimate.* London: Kaye & Ward, 1979, p. 174–180.
16. MEYER, M., P. SCHEID, G. RIEPL, H.-J. WAGNER, AND J. PIIPER. Pulmonary diffusing capacities for O_2 and CO measured by a rebreathing technique. *J. Appl. Physiol.: Respirat. Environ. Exercise Physiol.* 51: 1643–1650, 1981.
17. PIIPER, J., AND P. SCHEID. Model for capillary-alveolar equilibration with special reference to O_2 uptake in hypoxia. *Respir. Physiol.* 46: 193–208, 1981.
18. PUGH, L. G. C. E. Resting ventilation and alveolar air on Mount Everest: with remarks on the relation of barometric pressure to altitude in mountains. *J. Physiol. London* 135: 590–610, 1957.
19. PUGH, L. G. C. E. Cardiac output in muscular exercise at 5800 m (19,000 ft). *J. Appl. Physiol.* 19: 441–447, 1964.
20. PUGH, L. G. C. E., M. B. GILL, S. LAHIRI, J. S. MILLEDGE, M. P. WARD, AND J. B. WEST. Muscular exercise at great altitudes. *J. Appl. Physiol.* 19: 431–440, 1964.
21. RAHN, H., AND A. B. OTIS. Man's respiratory response during and after acclimatization to high altitude. *Am. J. Physiol.* 157: 445–449, 1946.
22. SIGGAARD-ANDERSEN, O. An acid-base chart for arterial blood with normal and pathophysiological reference areas. *Scand. J. Clin. Lab. Invest.* 27: 239–245, 1971.
23. SIGGAARD-ANDERSEN, O., AND K. ENGEL. A new acid-base nomogram. An improved method for the calculation of the relevant blood acid-base data. *Scand. J. Clin. Lab. Invest.* 12: 177–186, 1960.
24. WAGNER, P. D. Diffusion and chemical reaction in pulmonary gas exchange. *Physiol. Rev.* 57: 257–312, 1977.
25. WARREN, C. B. M. Alveolar air on Mount Everest. *J. Physiol. London* 96: 34P–35P, 1939.
26. WEST, J. B. Diffusing capacity of the lung for carbon monoxide at high altitude. *J. Appl. Physiol.* 17: 421–426, 1962.
27. WEST, J. B. American Medical Research Expedition to Everest, 1981. *Physiologist* 25: 36–38, 1982.
28. WEST, J. B., S. J. BOYER, D. J. GRABER, P. H. HACKETT, K. H. MARET, J. S. MILLEDGE, R. M. PETERS, JR., C. J. PIZZO, M. SAMAJA, F. H. SARNQUIST, R. B. SCHOENE, AND R. M. WINSLOW. Maximal exercise at extreme altitudes on Mount Everest. *J. Appl. Physiol.: Respirat. Environ. Exercise Physiol.* 55: 688–702, 1983.
29. WEST, J. B., P. H. HACKETT, K. H. MARET, J. S. MILLEDGE, R. M. PETERS, JR., C. J. PIZZO, AND R. M. WINSLOW. Pulmonary gas exchange on the summit of Mount Everest. *J. Appl. Physiol.: Respirat. Environ. Exercise Physiol.* 55: 678–687, 1983.
30. WEST, J. B., S. LAHIRI, K. H. MARET, R. M. PETERS, JR., AND C. J. PIZZO. Barometric pressures at extreme altitudes on Mt. Everest: physiological significance. *J. Appl. Physiol.: Respirat. Environ. Exercise Physiol.* 54: 1188–1194, 1983.
31. WEST, J. B., AND P. D. WAGNER. Predicted gas exchange on the summit of Mt. Everest. *Respir. Physiol.* 42: 1–16, 1980.
32. WINSLOW, R. M., M. SAMAJA, AND J. B. WEST. Red cell function at extreme altitude on Mount Everest. *J. Appl. Physiol.: Respirat. Environ. Exercise Physiol.* 56: 109–116, 1984.

2

Hypoxic Ventilatory Response and Exercise Ventilation at Sea Level and High Altitude

ROBERT B. SCHOENE

Division of Respiratory Disease, University of Washington, Harborview Medical Center, Seattle, Washington

THE VENTILATORY RESPONSE TO HYPOXIA (HVR) is an inborn characteristic that varies between individuals (43). Like so many physiological responses, its purpose is to maximize the chance of survival in a perilous environment. An adequate level of ventilation is necessary to maintain sufficient partial pressure of oxygen in alveolar gas and arterial blood ($P_{A_{O_2}}$ and Pa_{O_2}, respectively). For healthy individuals at sea level, maintaining ventilation is easy and is mediated largely by the partial pressure of carbon dioxide (P_{CO_2}), with hypoxia acting as a safeguard mechanism. In states of disease where gas exchange and oxygenation are impaired, however, HVR becomes increasingly important. Also at high altitude, where ambient P_{O_2} is precariously low, it plays an integral part in survival. This chapter discusses the relationship of HVR and exercise performance both at sea level and high altitude, in sojourners as well as in natives to high altitude. In addition, recent data pertinent to hypoxic drives and exercise performance collected on the 1981 American Medical Research Expedition to Everest are presented.

HYPOXIC VENTILATORY RESPONSE AND EXERCISE VENTILATION

Several studies have shown a correlation between resting HVR and exercise ventilation. Martin and co-workers (25) studied both hypoxic and hypercapnic (HCVR) drives and exercise ventilation in 16 athletes; at light, heavy, and maximal levels of exercise they found a significant correlation between HVR, HCVR, and ventilatory equivalent ($\dot{V}_E/\dot{V}_{CO_2}$) (Fig. 1). During light exercise, $\dot{V}_E/\dot{V}_{CO_2}$ had a significant negative correlation with Pa_{CO_2} (Fig. 2). This finding is important in relation to exercise at extreme altitude. In a subsequent experiment Martin et al. (24) studied endurance and nonendurance

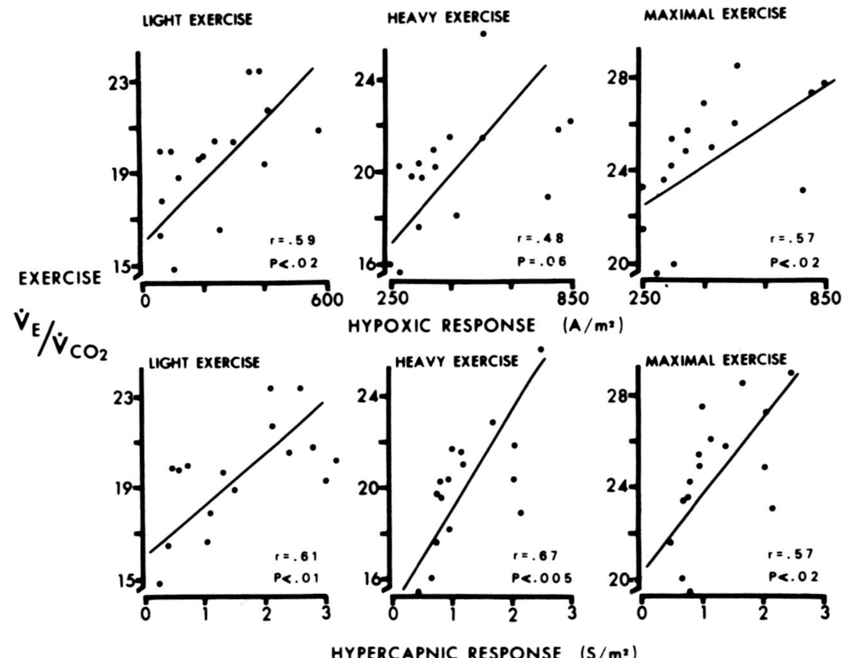

FIG. 1. Positive correlation of ventilation per unit metabolic rate ($\dot{V}E/\dot{V}CO_2$) during light, heavy, or maximal exercise with hypoxic and hypercapnic ventilatory responses. [From Martin et al. (25).]

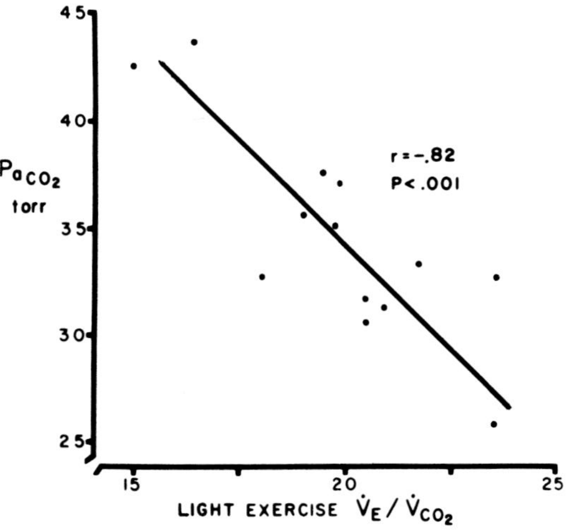

FIG. 2. Significant negative relationship between $\dot{V}E/\dot{V}CO_2$ and partial pressure of CO_2 in arterial blood (Pa_{CO_2}) during light exercise (same relationship found during heavy exercise). [From Martin et al. (25).]

athletes and nonathletes; they found low, moderate, and high chemosensitivity, respectively (see Fig. 3). Exercise ventilation corresponded with each group's ventilatory drives at light and heavy exercise levels.

Exogenous or endogenous manipulation of chemosensitivity can also substantiate the relationship between resting HVR and exercise hyperpnea. For instance, exogenous progesterone (medroxyprogesterone acetate) augments ventilatory drives (34, 38, 49). Exercise ventilation also increased correspondingly when medroxyprogesterone acetate was administered (31, 38). Higher ventilatory drives and exercise ventilation were also noted during the luteal phase of the menstrual cycle, when endogenous progesterone is high (34).

The carotid body is important in the mediation of the ventilatory response both to hypoxia (26) and exercise (48), and removing the carotid body significantly diminishes HVR, HCVR, and exercise hyperpnea (23, 41). Neural output from the carotid body can also be diminished significantly during exercise when inhalation of oxygen is undertaken (6, 8). This finding intimates that the peripheral chemosensor at least modulates exercise ventilation. Because Po_2 does not drop during exercise in normal subjects at sea level, other influences must act on the carotid body to stimulate exercise ventilation. The lactic acidosis that ensues during high levels of muscular work may stimulate the carotid body directly (2). An increase in sympathetic discharge and release of catecholamines may also reduce blood flow to the peripheral chemosensor and impair the clearance of the products of anaerobic metabolism (22, 28, 47). These effects can stimulate neural discharge from the carotid body and affect exercise hyperpnea.

When a physically active sojourner is exposed to high altitude, the carotid body probably helps to modulate the hyperpnea of hypoxia and exercise. Chronic exposure to hypoxia causes hypertrophy of the carotid body (1), although this enlargement does not generally result in brisker ventilatory drives. This relationship is discussed in HYPOXIC VENTILATORY RESPONSE AND HIGH ALTITUDE, p. 22.

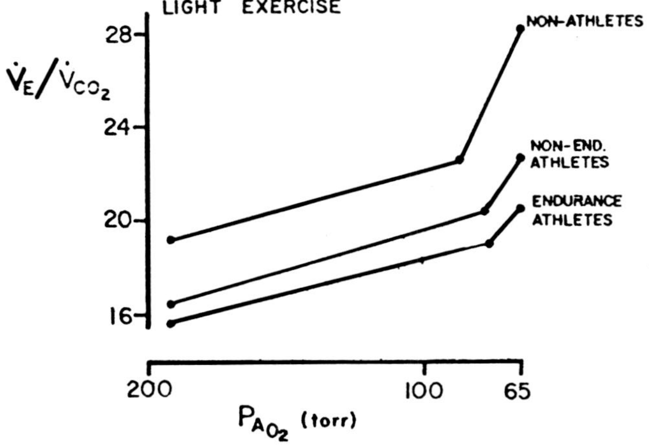

FIG. 3. Mean $\dot{V}E/\dot{V}CO_2$ at various partial pressures of O_2 in alveolar gas ($P_{A_{O_2}}$) during light exercise. Endurance athletes breathed significantly less per unit $\dot{V}CO_2$ than nonathletes at any $P_{A_{O_2}}$ ($P < 0.05$). [From Martin et al. (24).]

BLUNTED HYPOXIC VENTILATORY DRIVES

Blunted ventilatory drives have been associated with athletes of outstanding aerobic and endurance capabilities (3, 4, 24, 25, 34). Less exercise dyspnea and a lower work of breathing at given levels of work result from this blunted chemosensitivity and may make exercise not only more comfortable but also more efficient. This inborn characteristic may be one of several physiological and psychological factors, which if found in a favorable combination may predispose certain individuals to outstanding athletic performance. In three of these studies (24, 25, 34) individuals with blunted chemosensitivity had a lower ventilatory equivalent, higher $P_{A_{CO_2}}$, and lower $P_{A_{O_2}}$ during exercise (Fig. 3). At low altitudes, where a surfeit of oxygen is available, a lower work of breathing and lower dyspnea at high levels of exercise are probably an advantage in these trained endurance athletes. The slightly higher $P_{A_{CO_2}}$ and lower $P_{A_{O_2}}$ have no detrimental physiological effect on performance in low-altitude environments.

HYPOXIC VENTILATORY RESPONSE AND HIGH ALTITUDE

At high altitude, where the ambient P_{O_2} is low and where a diffusion limitation to oxygen widens the alveolar-arterial oxygen gradient even further (46), an adequate ventilatory response is necessary to maintain a sufficient alveolar and subsequent tissue oxygen level. It seems paradoxical, therefore, that there are populations who live at high altitude with blunted chemosensitivity. Lahiri et al. (19) and Milledge and Lahiri (27) documented low ventilatory drives and exercise ventilation in a few Sherpas, whereas other investigators have noted blunted chemosensitivity in high-altitude natives of the Andes (5, 18, 36, 42) and some inhabitants of Leadville, Colorado (17). In these two latter groups, there is a significant incidence of chronic mountain sickness, whereas in the Himalayan native this clinical finding is rare. Recent data from Hackett et al. (11) suggest that Sherpas may have a distribution of drives more similar to lowlanders. Whatever the situation, other compensatory mechanisms that improve gas exchange, such as an increase in diffusion capacity (7, 30), may compensate for relative hypoventilation so that individuals born and raised at altitude do not depend as much on a brisk HVR. Also an improved extraction of oxygen at the tissue level may be important; however, this possibility has not been investigated.

On the other hand, an adequate ventilatory response to altitude is important to the sojourner, and the course of its response has been well described (9, 29, 37). The stimulus of hypoxia during acute exposure to high altitude increases ventilation. The subsequent respiratory alkalosis suppresses this initial response slightly; as partial renal compensation for the alkalosis continues over several days, however, ventilation is increased even further.

Barcroft (3), Haldane (14), and more recently Hackett et al. (12), King and Robinson (16), Larson et al. (21), and Sutton et al. (39) have emphasized the relationship between performance at high altitude and sufficient ventilation. In fact a lower ventilatory response to altitude has been linked with recurrent pulmonary edema (15, 20) and acute (12, 16, 21, 39) and chronic (17) mountain sickness. Therefore respiratory stimulation with exogenous agents such as acetazolamide improves oxygen desaturation during sleep at altitude (17, 40), acute mountain sickness (13, 21), and chronic mountain sickness (40).

HYPOXIC VENTILATORY RESPONSE AND EXTREME ALTITUDE

At modest altitudes the barometric pressure of oxygen carries with it some margin of safety; however, at altitudes greater than 7,500 m, this margin is so low that an inadequate HVR could result in intolerably low alveolar oxygen and high carbon dioxide levels. Martin (Fig. 2; 25) has shown that, at low altitude, individuals with blunted chemosensitivity exercise with a lower ventilatory equivalent and higher $P_{A_{CO_2}}$. These few mmHg of P_{CO_2} in the alveolus would lower $P_{A_{O_2}}$. If these data can be extrapolated to high altitude, this difference in P_{O_2} may profoundly affect arterial and tissue oxygenation. West and Wagner (46) hypothesized that on the summit of Mount Everest (8,848 m) $P_{A_{O_2}}$ and $P_{A_{CO_2}}$ might be ~30 and ~10 mmHg, respectively. Actual data collected on the summit in one individual showed that an even greater ventilatory response than was predicted was found with a $P_{A_{CO_2}}$ of 7–8 mmHg (44).

The climber to extreme altitude may therefore be an endurance athlete with different characteristics from the elite long-distance runner at sea level. He or she has high aerobic capacity, muscular power, and endurance similar to the low-altitude athlete but, because of the requirements of the environment, may not have blunted ventilatory drives. Recently 14 individuals who had performed well without altitude illness at altitudes of 7,500 m or greater were studied (32). Their HVR values were measured and compared to 10 age-matched controls and 10 nationally or world-ranked distance runners. The group of climbers, which included eight summit climbers to Mount Everest or K^2, had significantly higher HVR values than did the controls or runners (Fig. 4). These data imply that *1)* climbers to extreme altitude may be athletes of high aerobic capacity but with chemosensitivity that is different from that of the elite distance runner; *2)* the price of dyspnea and increased work of breathing in these climbers may be far outweighed by the increased alveolar oxygen that is a result of their increased ventilation; *3)* because expedition climbing involves weeks or months at high altitude during rest, sleep, and exercise, this improved oxygenation may minimize the chronic manifestations of more severe hypoxemia. These data do not necessarily preclude climbers with blunted chemosensitivity from ascending to extreme altitude with supplemental oxygen. In addition this study does not confirm the relationship between HVR measured at sea level and exercise performance at altitude.

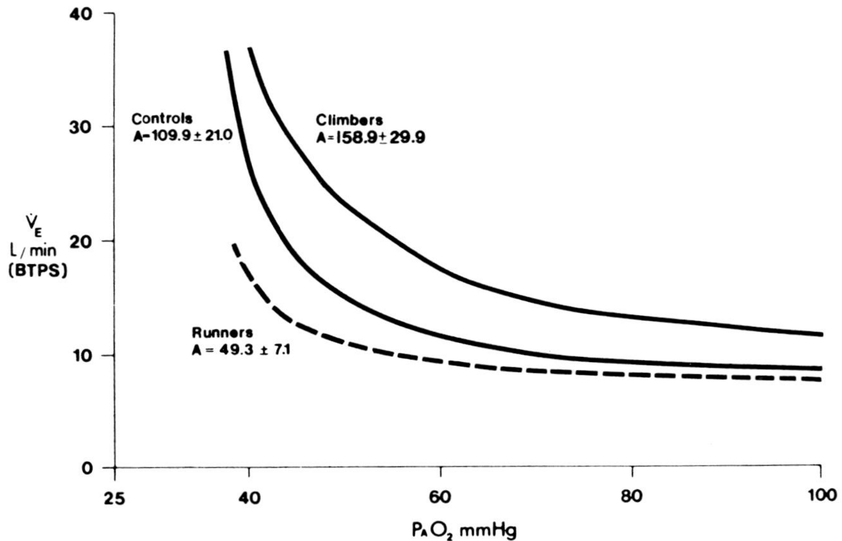

FIG. 4. Hypoxic ventilatory response plotted as minute ventilation ($\dot{V}E$) vs. $P_{A_{O_2}}$, showing hyperbolic relationship expressed as shape parameter *A*. Runners had significantly lower responses than climbers and controls. [From Schoene (32).]

AMERICAN MEDICAL RESEARCH EXPEDITION TO EVEREST

The 1981 American Medical Research Expedition to Everest offered the opportunity to evaluate the following questions: *1*) Does HVR at sea level bear any relationship to HVR in the same individual after acclimatization to high altitude? *2*) Is there a relationship between HVR and exercise ventilation at altitude? *3*) Does exercise ventilation affect oxygen desaturation during exercise in hypoxic environment? *4*) Is HVR related to performance at altitude?

At sea level the HVR of eight expedition members was measured [using isocapnic progressive hypoxia (43)] as the increase in ventilation from normoxia to a $P_{A_{O_2}}$ of 40 mmHg ($\Delta\dot{V}E_{40}$). At 5,400 m a technique based on transient breaths of nitrogen was utilized (10); HVR was calculated as the slope of $\dot{V}E$ over arterial oxygen saturation. The distribution of HVR is shown in Table 1. Acclimatization to moderate altitude did not significantly alter the relationship of HVR between individuals ($P < 0.05$). A few other individuals who could not be measured at sea level were measured at 5,800 m and categorized according to the vigor of their ventilatory drives.

Exercise testing was performed at sea level on a bicycle ergometer with subjects breathing a fractional concentration of oxygen in dry inspired gas ($F_{I_{O_2}}$) of 0.209 or, in two individuals, an $F_{I_{O_2}}$ of 0.10. With a similar protocol, testing was repeated after acclimatization to 6,300 m (barometric pressure = 350 mmHg) breathing an $F_{I_{O_2}}$ of 0.209 or, in two individuals, 0.14. Figure 5 compares exercise ventilation ($\dot{V}E$/kpm) at sea level and after acclimatization to 6,300 m. The stimuli of exercise and the hypoxia of altitude significantly increased the ventilatory equivalent ($\dot{V}E/\dot{V}O_2$) at comparable work loads.

Exercise ventilation ($\dot{V}E \cdot \dot{V}O_2^{-1} \cdot kpm^{-1}$) was then plotted according to the grouping of HVR, i.e., high, moderate, and low responders. The results, shown in Figure 6, indicate that, at low and moderate levels of exercise, HVR and exercise ventilation are related. At high levels of exercise this relationship is not consistent, possibly because other factors influence exercise ventilation above the anaerobic threshold. These factors probably mediate ventilation at the peripheral chemoreceptor (48).

Oxygen desaturation is well documented in healthy individuals during exercise at altitude (45). Marked differences in oxygen desaturation were noted

TABLE 1. *Hypoxic ventilatory responses at sea level and 5,400 m*

Subject	Hypoxic Ventilatory Response, liters/min	
	Sea level, $\dot{V}E_{40}$ BTPS	5,400 m, $\dot{V}E/Sa_{O_2}\%$
CK	33.9	0.8
CP	26.6	0.75
PH	12.7	0.7
DJ	11.7	0.6
SB	7.1	1.0
FS	7.0	0.55
JE	4.9	0.15
RS	3.2	0.22
JL		0.17

Correlation between hypoxic ventilatory response determinations is significant ($P < 0.05$). Subjects are grouped into high (*CK, CP*), moderate (*PH, DJ, SB, FS*), and low (*JE, RS, JL*) responders to hypoxia.

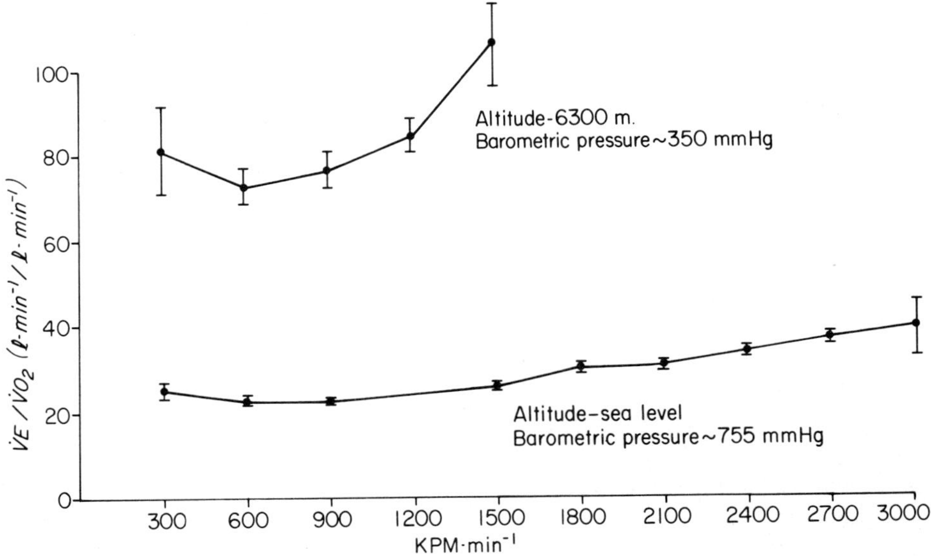

FIG. 5. Work load vs. ventilatory equivalent ($\dot{V}E/\dot{V}O_2$) of expedition members both at sea level (*lower curve*) and at 6,300 m (*upper curve*).

between the HVR groups (Fig. 6). A similar pattern between two individuals, one with high, the other with low HVR, was noted during exercise after acute exposure to a hypoxic gas mixture at both sea level (Fig. 7) and 6,300 m (Fig. 8, *bottom panel*). These differences between the groups may in part be secondary to the variation in ventilatory response to exercise so that the hyperventilators have relatively higher $P_{A_{O_2}}$ values and lower P_{CO_2} values and subsequently higher arterial oxygenation. In addition the individuals with a higher ventilation have a more marked alkalosis (44), probable left shift of the oxygen-hemoglobin–dissociation curve, and a subsequently higher arterial oxygen saturation. In the hypoventilators the relative alveolar hypoxia may decrease the pressure gradient for oxygen diffusion from the alveolus to the pulmonary capillary and subsequently to the tissues (46). Despite quite disparate arterial saturations, individuals were able to perform comparable work loads at similar

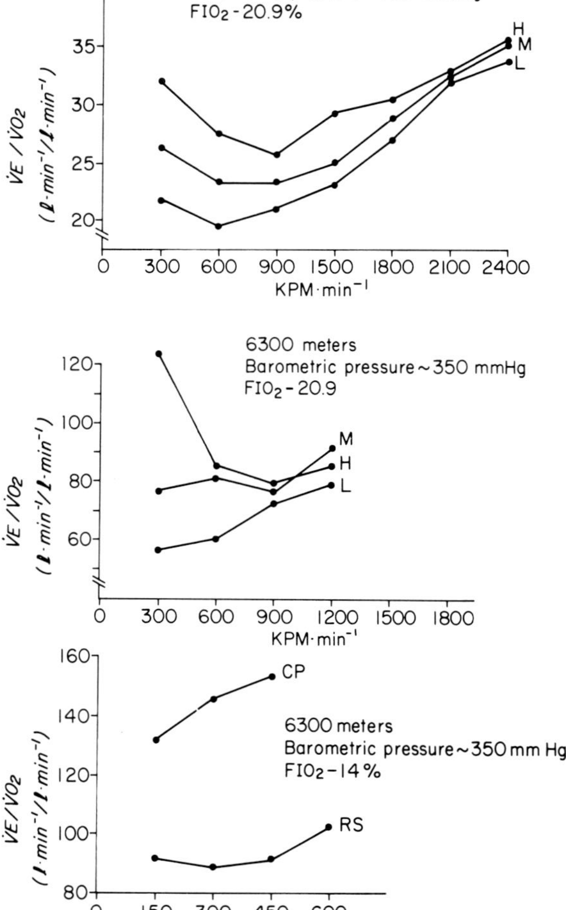

FIG. 6. Work load vs. $\dot{V}_E/\dot{V}_{O_2}$) of the high (H), moderate (M), and low (L) hypoxic ventilatory responders at sea level (*top panel*) and 6,300 m (*middle panel*). Subjects *CP* and *RS* are shown in *bottom panel* during exercise at 6,300 m, breathing a fractional concentration of O_2 in dry inspired gas ($F_{I_{O_2}}$) of 0.14.

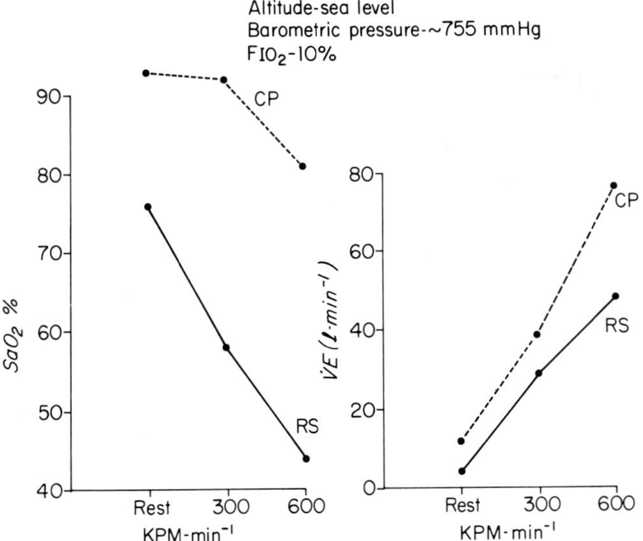

FIG. 7. Work load vs. Sa_{O_2} % (*left panel*) and $\dot{V}_E$ (*right panel*) in subjects *CP* and *RS* while breathing a hypoxic FI_{O_2} of 0.10.

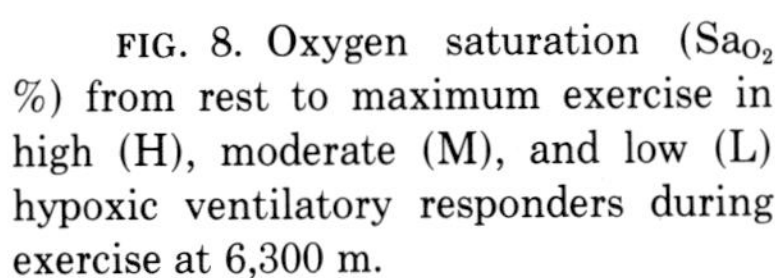

FIG. 8. Oxygen saturation (Sa_{O_2} %) from rest to maximum exercise in high (H), moderate (M), and low (L) hypoxic ventilatory responders during exercise at 6,300 m.

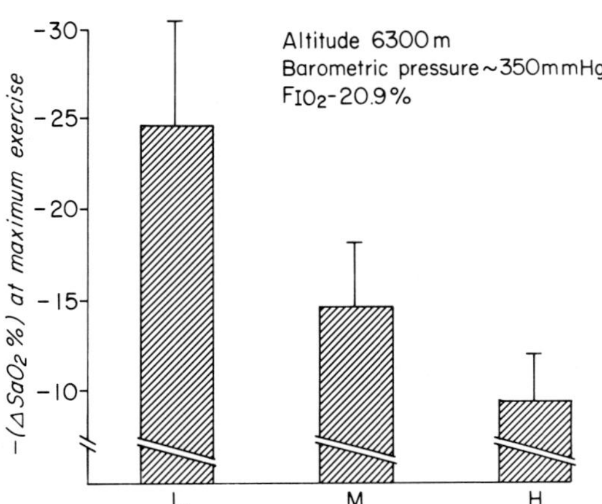

levels of oxygen consumption. How the individuals with lower oxygen saturations can extract similar amounts of oxygen with a smaller diffusion gradient to the tissues is not clear. This ability to consume comparable levels of oxygen at different arterial saturations may be secondary to differences in cardiac output or arteriovenous oxygen gradient.

Performance at altitude over the course of an expedition is difficult to measure except by using the maximum altitude attained or the highest altitude where one slept. Table 2 gives the altitudes attained by climbers during the Everest expedition; apparently the high and moderate HVR groups both climbed and slept higher than the low HVR groups. Admittedly these data

TABLE 2. *Highest altitude attained on American Medical Research Expedition to Everest*

Hypoxic Ventilatory Response Group	Highest Sleeping Altitude, m	Highest Climbing Altitude, m
High	8,050 ± 0	8,848 ± 0
Moderate	7,713 ± 403	8,250 ± 399
Low	7,017 ± 633	7,400 ± 260

Values are means ± SD.

need to be pursued, but they certainly suggest that HVR may be related to performance at extreme altitude.

SUMMARY

The HVR varies between individuals and correlates with exercise ventilation both at low and high altitudes. This relationship determines certain characteristics that may affect performance in both environments. If persons have other attributes necessary for exceptional performance, a blunted HVR may contribute to success at low altitude in middle- and long-distance athletic competition. On the other hand, at a very high altitude, where the P_{O_2} is low, a blunted HVR may not ensure an adequate $P_{A_{O_2}}$ for prolonged physical performance.

This paper has reviewed *1*) the relationship of HVR to exercise ventilation at sea level and high altitude; *2*) the association of blunted HVR with outstanding endurance athletes; *3*) the association of normal or high HVR with success at extreme altitude; and *4*) the data recently accumulated on the 1981 American Medical Research Expedition to Everest, which support the hypothesis that at high altitude HVR is related to exercise ventilation, protects oxygen saturation during exercise, and may contribute to other factors that determine whether one can sleep at or climb to extreme altitude.

In summary, data from the expedition to Everest show that HVR correlates with the response after acclimatization to altitude, predicts exercise ventilation at sea level and altitude, protects oxygen saturation during exercise, and may predict adaptation and performance at extreme altitude. Arterial oxygen desaturation during exercise at altitude, however, is only partly explained by ventilation. In addition a high HVR may be necessary for climbers to extreme altitude who do not use supplemental oxygen.

REFERENCES

1. ARIAS-STELLA, J., AND J. VALCAREL. Chief cell hyperplasia in the human carotid body at high altitude. *Hum. Pathol.* 7: 361–373, 1976.
2. ASTRAND, P., AND K. RODAHL. *Textbook of Workbook Physiology*. New York: McGraw-Hill, 1977.
3. BARCROFT, J. B. *The Respiratory Function of the Blood. Part I.* Cambridge, UK: Cambridge Univ. Press, 1925.
4. BYRNE-QUINN, E., J. V. WEIL, I. E. SODAL, G. F. FILLEY, AND R. F. GROVER. Ventilatory control in the athlete. *J. Appl. Physiol.* 30: 91–98, 1971.
5. CHIODI, H. Respiratory adaptations to chronic high altitude hypoxia. *J. Appl. Physiol.* 10: 81–87, 1957.
6. CUNNINGHAM, D. J. C. Integrative aspects of the regulation of breathing: a personal view. In: *Respiratory Physiology I*, edited by J. G. Widdicombe. Baltimore, MD: University Park, 1974, vol. 2, p. 303. (Int. Rev. Physiol.

Ser.)

7. DEGRAFF, A. C., JR, R. F. GROVER, R. L. JOHNSON, JR., J. W. HAMMOND, JR., AND J. M. MILLER. Diffusing capacity of the lung in Caucasians native to 3,100 m. *J. Appl. Physiol.* 29: 71–76, 1970.
8. DEJOURS, P. Control of respiration in muscular exercise. In: *Handbook of Physiology. Respiration*, edited by W. O. Fenn and H. Rahn. Washington, DC: Am. Physiol. Soc., 1964, sect. 3, vol. I, chapt. 25, p. 631–648.
9. DEMPSEY, J. A., AND H. V. FORSTER. Mediation of ventilatory adaptations. *Physiol. Rev.* 62: 262–346, 1982.
10. EDELMAN, N. H., P. E. EPSTEIN, S. LAHIRI, AND N. S. CHERNIACK. Ventilatory responses to transient hypoxia and hypercapnia in man. *Respir. Physiol.* 17: 302–314, 1973.
11. HACKETT, P. H., J. T. REEVES, C. D. REEVES, R. F. GROVER, AND D. RENNIE. Control of breathing in Sherpas at low and high altitude. *J. Appl. Physiol.: Respirat. Environ. Exercise Physiol.* 49: 374–379, 1980.
12. HACKETT, P. H., D. RENNIE, S. E. HOFMEISTER, R. F. GROVER, E. G. GROVER, AND J. T. REEVES. Fluid retention in acute mountain sickness. *Respiration* 43: 321–329, 1982.
13. HACKETT, P. H., D. RENNIE, AND H. D. LEVINE. The incidence, importance, and prophylaxis of acute mountain sickness. *Lancet* 2: 1149–1154, 1976.
14. HALDANE, J. S. *Respiration.* New Haven, CT: Yale Univ. Press, 1982, p. 374.
15. HYERS, T. M., C. H. SCOGGIN, D. H. WILL, R. F. GROVER, AND J. T. REEVES. Accentuated hypoxemia at high altitude in subjects susceptible to high-altitude pulmonary edema. *J. Appl. Physiol.: Respirat. Environ. Exercise Physiol.* 46: 41–46, 1979.
16. KING, A. B., AND S. M. ROBINSON. Ventilation response to hypoxia and acute mountain sickness. *Aviat. Space Environ. Med.* 43: 419–421, 1972.
17. KRYGER, M., R. E. McCULLOUGH, D. COLLINS, C. H. SCOGGINS, J. V. WEIL, AND R. F. GROVER. Treatment of excessive polycythemia of high altitude with respiratory stimulant drugs. *Am. Rev. Respir. Dis.* 117: 455–464, 1978.
18. LAHIRI, S., F. F. KAO, T. VELASQUEZ, C. MARTING, AND W. PEZZIA. Irreversible blunted respiratory sensitivity to hypoxia in high altitude natives. *Respir. Physiol.* 6: 360–374, 1969.
19. LAHIRI, S., F. F. KAO, T. VELASQUEZ, C. MARTING, AND W. PEZZIA. Respiration of man during exercise at high altitude: highlander vs. lowlander. *Respir. Physiol.* 8: 361–375, 1970.
20. LAKSHMINARAYAN, S., AND D. J. PIERSON. Recurrent high altitude pulmonary edema with blunted chemosensitivity. *Am. Rev. Respir. Dis.* 111: 869–872, 1975.
21. LARSON, E. B., R. C. ROACH, R. B. SCHOENE, AND T. F. HORNBEIN. Acute mountain sickness and acetazolamide. *J. Am. Med. Assoc.* 248: 328–332, 1982.
22. LEE, K. D., R. A. MAYOU, AND R. W. TORRANCE. The effect of blood pressure upon chemoreceptor discharge to hypoxia and the modifications of this effect by the sympathetic adrenal system. *J. Quant. Exp. Physiol.* 49: 171, 1964.
23. LUGLIANI, R., B. J. WHIPP, C. SEARS, AND K. WASSERMAN. Effect of bilateral carotid body resection on ventilatory control at rest and during exercise in man. *N. Engl. J. Med.* 285: 1105–1111, 1971.
24. MARTIN, B. J., K. E. SPARKS, C. W. ZWILLICH, AND J. V. WEIL. Low exercise ventilation in endurance athletes. *Med. Sci. Sports* 11: 181–185, 1979.
25. MARTIN, B. J., J. V. WEIL, K. E. SPARKS, R. E. McCULLOUGH, AND R. F. GROVER. Exercise ventilation correlates positively with ventilatory chemoresponsiveness. *J. Appl. Physiol.: Respirat. Environ. Exercise Physiol.* 45: 557–564, 1978.
26. McDONALD, D. M. Peripheral chemoreceptors: structure-function relationships of the carotid body. In: *Regulation of Breathing*, edited by T. F. Hornbein. New York: Dekker, 1981, chapt. 3, p. 105–319.
27. MILLEDGE, J. S., AND S. LAHIRI. Respiratory control in lowlanders and sherpa highlanders at high altitude. *Respir. Physiol.* 2: 310–322, 1967.
28. NEIL, E., AND N. JOELS. The carotid glomus sensory mechanism. In: *Regulation of Human Respiration*, edited by D. J. C. Cunningham and B. B. Lloyd. Oxford, UK: Blackwell, 1963, p. 163.
29. RAHN, H., AND A. B. OTIS. Man's respiratory response during and after acclimatization to high altitude. *Am. J. Physiol.* 157: 445–462, 1949.
30. REMMERS, J. E., AND J. C. MITHOEFER. The carbon monoxide diffusing capacity in permanent residents at high altitude. *Respir. Physiol.* 6: 233–244, 1969.
31. ROBERTSON, H. T., R. B. SCHOENE, AND D. J. PIERSON. Effect of medroxyprogesterone acetate on exercise ventilation. *Clin. Physiol.* 2: 269–276, 1982.
32. SCHOENE, R. B. Control of ventilation in climbers to extreme altitude. *J. Appl. Physiol.: Respirat. Environ. Exercise Physiol.* 53: 886–890, 1982.
33. SCHOENE, R. B., D. J. PIERSON, H. T. ROBERTSON, S. LAKSHMINARAYAN, D. SHRADER, AND J. BUTLER. Effect of medroxyprogesterone acetate on respiratory drives and occlusion pressure. *Bull. Eur. Physiopathol. Respir.* 16: 645–653, 1980.
34. SCHOENE, R. B., H. T. ROBERTSON, D. J. PIERSON, AND A. P. PETERSON. Respiratory drives and exercise in menstrual cycles in athletic and nonathletic women. *J. Appl. Physiol.: Respirat. Environ. Exercise Physiol.* 50: 1300–1305, 1981.
36. SEVERINGHAUS, J. W., C. R. BAINTON, AND A. CARCELÉN. Respiratory insensitivity to hypoxia in chronically hypoxic man. *Respir. Physiol.* 1: 308–334, 1966.
37. SEVERINGHAUS, J. W., R. A. MITCHELL, B. W. RICHARDSON, AND M. M. SINGER. Respiratory control at high altitude suggesting active transport regulation of CSF pH. *J. Appl. Physiol.* 18: 1155–1166, 1963.
38. SKATRUD, J. B., J. A. DEMPSEY, AND D. J. KAISER. Ventilatory response to medroxyprogesterone acetate in normal subjects: time course and mechanism. *J. Appl. Physiol.: Respirat. Environ. Exercise Physiol.* 44: 939–944, 1978.
39. SUTTON, J. R., A. C. BRYAN, G. W. GRAY, E. S. HORTON, A. S. REBUCK, W. WOODLEY, I. D. RENNIE, AND C. S. HOUSTON. Pulmonary gas exchange in acute mountain sickness. *Aviat. Space Environ. Med.* 47: 1032–1037, 1976.
40. SUTTON, J. R., C. S. HOUSTON, A. L. MANSELL, M. D. McFADDEN, P. H. HACKETT, J. R. A. RIGG, AND A. C. P. POWLES. Effect of acetazolamide on hypoxemia during sleep at high altitude. *N. Engl. J. Med.* 301: 1329–1331, 1979.
41. WASSERMAN, K., B. J. WHIPP, S. N. KOYAL, AND M. G. CLEARY. Effect of carotid body resection on ventilatory and acid-base control during exercise. *J. Appl. Physiol.* 39: 354–358, 1975.
42. WEIL, J. V., E. BYRNE-QUINN, I. E. SODAL, G. F. FILLEY, AND R. F. GROVER. Acquired attenuation of chemoreceptor function in chronically hypoxic man at high altitude. *J. Clin. Invest.* 50: 186–195, 1971.
43. WEIL, J. V., E. BYRNE-QUINN, I. E. SODAL, W. O. FRIESSEN, B. UNDERHILL, G. F. FILLEY, AND R. F. GROVER. Hypoxic ventilatory response in normal

man. *J. Clin. Invest.* 49: 1061–1072, 1970.

44. WEST, J. B., P. H. HACKETT, K. H. MARET, J. S. MILLEDGE, R. M. PETERS, JR., C. J. PIZZO, AND R. M. WINSLOW. Pulmonary gas exchange on the summit of Mount Everest. *J. Appl. Physiol.: Respirat. Environ. Exercise Physiol.* 55: 678–687, 1983.
45. WEST, J. B., S. LAHIRI, M. B. GILL, J. S. MILLEDGE, L. G. C. E. PUGH, AND M. P. WARD. Arterial oxygen saturation during exercise at high altitude. *J. Appl. Physiol.* 17: 617–621, 1962.
46. WEST, J. B., AND P. D. WAGNER. Predicted gas exchange on the summit of Mt. Everest. *Respir. Physiol.* 42: 1–46, 1980.
47. WHALEN, W. J., AND P. NAIR. Some factors affecting tissue Po_2 in the carotid body. *J. Appl. Physiol.* 39: 562–566, 1975.
48. WHIPP, B. J., AND J. A. DAVID. Peripheral chemoreceptors and exercise hyperpnea. *Med. Sci. Sports* 11: 204–212, 1979.
49. ZWILLICH, G. W., M. R. NATILINO, F. D. SUTTON, AND J. V. WEIL. Effects of progesterone on chemosensitivity in normal man. *J. Lab. Clin. Med.* 92: 262–269, 1978.

3

Human Cerebral Function at Extreme Altitude

BRENDA D. TOWNES, THOMAS F. HORNBEIN, ROBERT B. SCHOENE, FRANK H. SARNQUIST, AND IGOR GRANT

Departments of Psychiatry and Behavioral Sciences, Anesthesiology, and Medicine, University of Washington, Seattle, Washington; Department of Anesthesiology, Stanford University Medical Center, Stanford, California; and Psychiatry Service, San Diego Veterans Administration and Medical Center, Department of Psychiatry, University of California, San Diego, La Jolla, California

IN THE FALL OF 1981 the American Medical Research Expedition to Everest completed a series of physiological and psychological studies on mountaineers ascending to the summit of Mount Everest. This expedition afforded the unique opportunity to observe the consequences of extreme, sustained hypoxia on human cerebral function. The goal was to ascertain whether exposing healthy acclimatized individuals to extreme high altitude causes long-term alterations in cognition or behavior indicative of hypoxic brain dysfunction.

The effect of acute hypoxia on cognition has been studied in the laboratory under simulated conditions. These investigations suggest impaired sensory, perceptual, and motor performance at altitudes to 6,100 m (3, 7, 10). Although mild hypoxia (to 2,314 m) may improve performance on simple motor tasks, more time is required to learn a new task at 3,048 m. The mountaineer therefore might perform routine, well-practiced tasks adequately but be impaired in unpracticed emergency conditions (5).

Naturalistic observations of Alpine mountaineers suggest an increasing impairment in sensory, motor, and complex cognitive abilities as a function of severity of hypoxia with increasing altitude. At the highest altitudes, the mountaineer may behave similar to an individual with a known acute organic brain syndrome (9).

How do increasing degrees of oxygen deprivation together with extreme climatic conditions produce these cognitive and behavioral changes? Selvamurthy et al. (15) made electroencephalographic recordings of 10 high-altitude native soldiers and 10 lowlander soldiers at sea level and at 3,500 m. Compared to native highlanders, lowlanders showed an increase in α-activity during acclimatization, which suggests cortical depression. This change was associated with lethargic behavior attributed to cerebral hypoxia. Forster et al. (6) recorded the electroencephalographic and visually evoked responses of seven healthy male subjects at sea level and for 12 successive days at 4,300 m altitude. They noted no changes in cerebral electrical activity in the first 2–3 h of hypoxia. Two subjects showed electrical changes during the first 4 days, suggesting cortical depression. All remaining subjects showed electrical changes after the 5th day, indicating cortical excitation. Anorexia, insomnia, irritability, increased ventilation, and depression occurred simultaneously. This suggests that cognitive and behavioral changes during hypoxia at high altitudes were related to measurable alterations in central nervous system function. Additionally individuals differ during the initial stages of acclimatization in both the type and rate of change in cerebral electrical activity.

The physiological, cognitive, and behavioral changes at high altitudes are presumably caused by alteration in oxygenation of the tissues, secondary either to reduced supply or slower utilization (1). In acute hypoxia a reduction of arterial blood saturation to 85% decreases capacity for mental concentration and abolishes fine muscular coordination. A reduction to 75% leads to faulty judgment, emotional lability, and impaired muscular function (18).

Clark et al. (4) investigated the permanence of cognitive and behavioral changes observed at high altitudes after the hypoxic episode is resolved. They tested 22 mountaineers prior to and 16–221 days after Himalayan climbs above 5,100 m with an extensive battery of psychological and neuropsychological measures (Wechsler Adult Intelligence Scale, Halstead-Reitan Neuropsychological Test Battery, etc.). They found no evidence of permanent cerebral dysfunction due to altitude exposure. By contrast Sharma et al. (16, 17) found that lowlanders required to live for 10 mo at high altitudes initially experience impairment of both motor coordination and speed; while the former resolved within 1 yr, the latter persisted over a 2-yr period. Ryn (14) studied a group of 20 male and 10 female Polish Alpinists during and "for several weeks" after a Himalayan expedition. Only the male climbers ascended over 5,500 m. Half of these experienced symptoms similar to an acute organic brain syndrome; "for several weeks after the expedition they continued to feel poorly, showing signs of apathy and abulia, and impaired memory." Additionally 11 of the 30 climbers (6 men, 5 women) had abnormal electroencephalograms immediately after the climb. Psychological testing (Bender Gestalt and Graham-Kendall) showed normal visual motor performance in only 13 and borderline in 12 and was suggestive of organic pathology in 5 climbers. Although no preascent measures were reported, these data imply individual differences in the degree to which cognitive, behavioral, and central nervous system disturbances persist after a prolonged hypoxic episode. Whether or not such changes are permanent was a central question of our study.

METHODS

Subjects

Subjects were the 21 members of the 1981 American Medical Research Expedition to Everest. All were males between 25 and 52 yr, with a mean age of 36.4 yr. Fifteen had MDs or PhDs.

Procedures

Prior to the expedition the following psychological tests were administered to subjects at the Neuropsychology Laboratory, San Diego Veterans Administration Hospital: Halstead-Reitan Battery (9), Repeatable Cognitive-Perceptual-Motor Battery (8), Selective Reminding Test (2), and the Wechsler Memory Scale (14).

A few tests were administered during the ascent because they required minimal equipment and were easily administered. These were the Selective Reminding Test, Finger Tapping Test, and B Trails from the Halstead-Reitan Battery and the Digit Symbol, Digit Vigilance, Visual Search, and Peg-Board Tests from the Repeatable Battery. Pretest measures were readministered in Katmandu after descent from the mountain. At an expedition meeting held in Colorado 1 yr later, the following tests were readministered: Halstead-Wepman Aphasia Screening Test, B Trails and the Finger Tapping Test from the Halstead-Reitan Battery, the Digit Vigilance Task from the Repeatable Battery, and a verbal passage from the Wechsler Memory Scale.

Seventeen subjects participated in the pretest 2 mo before the expedition and again at Base Camp (5,700 m) within 3.3 days of arrival (0–6 days). Sixteen of these subjects were tested at the Laboratory Camp (6,300 m) within 6.4 days of arrival (1–18 days), and 9 subjects were tested at Laboratory Camp after returning from the South Col (8,050 m) within 1.7 days of arrival (0–4 days). Posttesting occurred within 3 days of arrival in Katmandu for 19 subjects; 2 subjects were tested in Seattle, Washington, within the 2 mo after the expedition. Fifteen subjects were tested at the 11-mo follow-up examination. Data were analyzed with the Wilcoxon signed-rank test to compare differences in performance between the testing periods.

RESULTS

Table 1 summarizes the significant changes found between preexpedition, postexpedition, and follow-up performance on the neuropsychological tests. Out of 21 comparisons made between performances prior to the expedition and in Katmandu, 9 were statistically significant. This is greater than would be expected by chance alone. Performance improved between the pre- and postexpedition periods on tests of complex problem solving, including spatial problem solving (Tactual Performance Test) and abstract reasoning (Category Test). This improvement was due to practice.

Verbal learning and memory declined significantly between pretest and postexpedition testing in Katmandu as measured by the Wechsler Memory Scale. With Heaton's modification of this test, a verbal passage containing 24 items of information was read to the subject, who then recalled as much information as possible. The passage was repeated until 14 items were recalled or 5 trials completed, whichever came first. Subjects were asked to repeat as much of the paragraph as possible 30 min later. Scores were the number of items recalled on the last trial (short-term verbal memory), number of trials

TABLE 1. *Wilcoxon signed-rank tests comparing performance before, immediately after (in Katmandu), and 1 yr after expedition to Mount Everest*

	Results, Means ± SE			Paired Responses, z Values		
	Before	After	Follow-up	Before and after	After and follow-up	Before and follow-up
Improved Performance						
Tactual Performance Test (right hand)	4.68 ± 1.56	3.86 ± 1.46		2.72*		
Category Test	24.29 ± 15.46	11.05 ± 8.39		3.48**		
Decline in Performance						
Finger Tapping Test						
Right hand	53.71 ± 4.07	45.40 ± 6.18	48.40 ± 6.60	3.39**	1.32	2.20†
Left hand	47.65 ± 4.60	42.25 ± 5.96	41.73 ± 5.23	2.30†	0.66	2.93*
Criterion right	1.00 ± 0	0.14 ± 0.36	0.27 ± 0.46	3.06*	0.73	2.67*
Criterion left	1.00 ± 0	0.14 ± 0.36	0.13 ± 0.35	2.93†	0.54	2.93*
Wechsler Memory Scale						
Short-term verbal recall	18.12 ± 1.90	15.90 ± 2.15	17.13 ± 2.20	2.60*	2.12†	0.98
Trials to Criterion	1.24 ± 0.44	2.40 ± 1.54	2.27 ± 0.70	2.37†	0	2.67*
Long-term verbal recall	16.35 ± 2.91	12.70 ± 3.78	14.50 ± 2.85	2.32†	2.75	0.94
Aphasia Screening Test	0.59 ± 0.79	1.25 ± 1.25	0.47 ± 0.52	2.22†	2.31†	0.47

* $P < 0.01$; ** $P < 0.001$; † $P < 0.05$.

to criterion, and the number of items recalled 30 min later (long-term verbal memory). As can be seen in Table 1, subjects in Katmandu compared to preexpedition testing took longer to learn the passage to criterion, recalled fewer items immediately, and showed decay in long-term verbal memory. As passages were not counterbalanced and different examiners were used at the two evaluation periods, these results are tentative and require replication. The observed decrements in verbal memory were transient (preexpedition levels returned 1 yr later).

On the Halstead-Wepman Aphasia Screening Test the number of expressive language errors increased significantly between pretest and posttest in Katmandu (see Table 2). Except for one in multiplication, errors in reading, writing, spelling, etc., were not present prior to the ascent. Such errors furthermore are not expected among young, healthy, and intelligent subjects. Pearson product-moment correlations were computed with the highest altitude attained and change between pre- and postexpedition performance. Altitude attained was significantly related to an increase in aphasic errors ($r = 0.55$; $P < 0.02$). These expressive language difficulties were transient (subjects were functioning at preexpedition levels 1 yr later).

Finger-tapping speed decreased significantly over the course of the expedition (Table 1). Mean taps for the right hand were 53.7 (pretest), 52.6 (Base

TABLE 2. *Types of aphasia errors at posttest in Katmandu*

Type of Error	Item	Response
Reading	7 six 2	6–7–6–2
Writing	warning	warninG
Calculation	17 × 3	49, 52, 57
Spelling	triangle	trangle
Pronunciation	Massachusetts, Episcopal	Massachusess, Massachutetts, Ekpiscopal
Confusion of body parts	Place left hand to right ear	Right hand placed to right ear

Camp), 50.8 (Laboratory Camp), 48.1 (return from South Col), and 45.4 (Katmandu); mean taps for the left hand were 47.6, 46.1, 47.4, 45.1, and 42.2, respectively. The standard method of administering the Finger Tapping Test is to obtain five trials (10 s each) on each hand with no greater difference than five taps between trials. Before the expedition all subjects reached criterion. At Katmandu 15 of 20 subjects could not sustain motor speed and 13 of 16 subjects could not do so 1 yr later. Given 1 min rest, motor speed improved but would again decline. Immediately after the expedition and up to 1 yr later, rapid muscle fatigue was present bilaterally.

DISCUSSION

In this young and highly educated group of subjects using supplemental oxygen to climb Mount Everest, transient and long-lasting neurobehavioral effects were found after exposure to the extreme hypoxemia of high altitude. Transient effects included a mild deterioration in the learning, memory, and expression of verbal material. These impairments were present within 3 days of descent into Katmandu but not 1 yr later. A bilateral reduction in motor speed characterized by rapid muscle fatigue persisted 1 yr after completion of the study. Clark et al. (4) tested subjects 16 or more days after descent to sea level with no observed impairment over pretest performance. As in this study Ryn (14) found signs of memory impairment "for several weeks after the expedition."

Our findings with those of Ryn support at least transient decrements in verbal memory and expression. Possibly the hippocampus or temporal areas of the brain produced the observed decrements in verbal memory. This should be tested in the future by separating learning from the retrieval of information across visual, auditory, and tactile modalities.

Because prolonged motor impairments have been found by Sharma et al. (16, 17) as well as in our studies, the finding appears reliable. One hypothesis is that cerebellar functions are negatively affected by prolonged exposure to hypoxia at altitude. Alternately, motor cortex functions may be impaired. We expect to test these hypotheses in the future by examining motor functions in more detail (8, 10).

The authors wish to thank Robert Reed and Lucy Brysh of the San Diego Veterans Administration and Medical Center for their work in obtaining base-line testing.

Portions of this paper were presented at the meeting of the American Physiological Society, San Diego, California, October 12–14, 1982.

This work was supported by grants from the National Geographical Society; the United States Army, Department of Environmental Medicine; National Heart, Lung, and Blood Institute Clinical Investigator Grant HC-00906; and an American Lung Association Trudeau Scholar Award.

REFERENCES

1. BRIERLY, J. B. Cerebral hypoxia. In: *Greenfield's Neuropathy*, edited by W. Blackwood and J. A. N. Corsellis. London: Armand, 1977, p. 43–85.
2. BUSCHKE, H. Selective reminding for analysis of memory and learning. *J. Verb. Learn. Verb. Behav.* 13: 543, 1973.
3. CAHOON, F. L. Simple decision making at high altitude. *Ergonomics* 14: 157–164, 1972.
4. CLARK, C. F., R. K. HEATON, AND A. N. WIENS. Neuropsychological functioning after prolonged high altitude exposure in mountaineering. *Aviat. Space Environ. Med.*, in press.
5. ERNSTING, J. Prevention of hypoxia—acceptable compromises. *Aviat. Space Environ. Med.* 49: 495–502, 1978.
6. FORSTER, H. V., R. J. SOTO, J. A. DEMPSEY, AND M. J. HOSKO. Effect of sojourn at 4,300 m altitude on electroencephalogram and visual evoked response. *J. Appl. Physiol.* 39: 109–113, 1975.
7. KOBRICK, J. L. Effects of hypoxia on peripheral visual response to dim stimuli. *Percept. Mot. Skills* 41: 467–474, 1975.
8. LEWIS, R. F., AND P. M. RENNICK. *Manual for the Repeatable Cognitive-Perceptual-Motor Battery.* Grosse Pointe Park, MI: Axon, 1979.
9. McFARLAND, R. A. Psychophysiological studies at high altitudes in the Andes. *J. Comp. Psychol.* 24: 189–220, 1937.
10. McFARLAND, R. A. Psychophysiological implications of life at altitude and including the role of oxygen in the process of aging. In: *Physiological Adaptation: Desert and Mountain*, edited by M. K. Yousef, S. M. Horvath, and R. W. Bullard. New York: Academic, 1972.
11. POTVIN, A. R., W. W. TOURTELLOTE, W. G. HENDERSON, AND D. N. SNYDER. Quantitative examination of neurological function: reliability and learning effects. *Arch. Phys. Med. Rehabil.* 56: 438–442, 1975.
12. REITAN, R. M., AND L. A. DAVISON (editors). *Clinical Neuropsychology: Current Status and Applications.* Washington, DC: Winston, 1974.
13. RUSSELL, E. W. A multiple scoring method for the assessment of complex memory functions. *J. Consult. Clin. Psychol.* 43: 800–809, 1975.
14. RYN, Z. Psychopathology in alpinism. *Acta Med. Pol.* 12: 453–467, 1976.
15. SELVAMURTHY, W., R. K. SAXENA, N. KRISHNAMURTHY, M. L. SURI, AND M. S. MALHOTRA. Changes in EEG pattern during acclimatization to high altitude (3,500 m) in man. *Aviat. Space Environ. Med.* 49: 968–971, 1978.
16. SHARMA, V. M., AND M. S. MALHOTRA. Ethnic variations in psychological performance under altitude stress. *Aviat. Space Environ. Med.* 47: 248–251, 1976.
17. SHARMA, V. M., M. S. MALHOTRA, AND A. S. BASKARAN. Variations in psychomotor efficiency during prolonged stay at high altitude. *Ergonomics* 18: 511–516, 1975.
18. WARD, M. *Mountain Medicine. A Clinical Study of Cold and High Altitude.* London: Crosby, Lockwood, Staples, 1975.

4

Metabolic and Endocrine Changes at Altitude

F. DUANE BLUME

Center for Physiological Research, California State College, Bakersfield, Bakersfield, California

A UNIQUE OPPORTUNITY to study the effects of prolonged exposures to altitudes above 5,000 m was provided by the 1981 American Medical Research Expedition to Everest (AMREE). A detailed assessment of the fasting metabolic state, in which the levels of key metabolites and hormones were measured, was included in the program. Fourteen adult male subjects between the ages of 28 and 52 yr (mean = 36) were studied at sea level and again during exposures to altitudes of 5,400 m and 6,300 m. This chapter details the similarities and differences between previous studies at moderate altitudes and the AMREE findings at more extreme altitudes.

BODY COMPOSITION

One of the most common observations made on sojourners at high altitude is an initial loss of body weight. Earlier studies showed that the weight loss at moderate altitudes stabilizes in adults after several weeks (5–8, 35, 37). Only at extreme altitudes (above 5,000 m) does the weight loss appear to be progressive and continuous (28).

The initial weight loss has been attributed to a combination of effects, namely, reduced dietary intake (7, 8, 35), increased water loss mainly from the lungs due to hyperventilation (6), and the loss of stored body fat (5, 37). The reduction in dietary intake resulting from the anorexia of acute mountain sickness limits the amount of exogenous energy available to the body, which forces the tissues to utilize stored energy materials. It is not surprising that several metabolic changes have been observed at altitude, including the mobilization of body fat (18, 37) and the development of a negative nitrogen balance (8, 12). Acclimatization ameliorates the symptoms of acute mountain sickness, but the dependence on fat persists (38).

High-altitude mountaineering at altitudes above 5,000 m has revealed other dimensions of the body's responses to hypoxia. Most climbers experience a greater loss of body weight at high altitudes than at lower elevations. It is not clear to what extent this loss is due to increased climbing activity, to limited food availability, or to more complex physiological adjustments that cause anorexia or influence food preferences. Many climbers claim to experience a preference for sweets at altitude, which has led expedition planners to provide high-carbohydrate diets on the mountain. This shift may result from internal adjustments requiring metabolic fuels that most efficiently use oxygen for combustion. Despite a change in diet that optimizes metabolism in this way, each climber seems to have a maximum altitude to which he can ascend before weight loss is both inevitable and ongoing. This suggests that extreme hypoxia produces profound metabolic effects that are exacerbated with time.

Team members in the AMREE study lost an average of 1.9 kg of body weight during the 250-km, 23-day approach march to Base Camp (5,400 m). Nearly 75% of the loss could be attributed to the reduction in body fat stores and was directly correlated with the initial amount of body fat. During subsequent exposures to altitudes higher than 5,400 m averaging 26 days, the weight loss increased markedly for the group (4.0 kg). However, body fat accounted for less than 50% of the loss at these altitudes. Substantial reductions in limb circumferences during this period suggested that muscle protein was being catabolized, contributing significantly to the weight change. Surprisingly the inferred changes in muscle mass occurred long before fat stores were exhausted. Thus at extreme altitudes increased gluconeogenesis seems to be the result of decreased dietary intake and probably some other metabolic alterations.

The difference between short-term changes in body weight observed at moderate altitudes and the continuous loss at extreme altitudes is not fully understood. At extreme altitudes the demand for energy sources apparently exceeds the supply from exogenous sources, resulting in the utilization of stored fat and protein. As a group, subjects involved in strenuous climbing activity lost more weight than those who were not. To what extent reductions in dietary intake contributed to the weight loss on Mount Everest is not known because food intake was difficult to monitor consistently because of the varied climbing activities. Limited data were taken on four subjects over 3-day periods at sea level and at 6,300 m. Caloric intake decreased 25% at altitude. Carbohydrates represented 42% of the caloric intake at sea level as compared to 52% at altitude. This modest shift probably resulted from the great availability of carbohydrates at altitude rather than a noticeable change in preference.

Changes in dietary intake notwithstanding, earlier studies provided evidence that conditions of hypoxemia affect intestinal absorption of foodstuffs. Pugh (28) reported evidence of malabsorption of fat in long-term exposure to 5,500 m, and Milledge (22) reported decreased xylose absorption in patients suffering from a variety of chronic conditions causing low arterial oxygen saturations. During AMREE significant reductions in both fat and xylose absorption of 48% and 24%, respectively, were observed during stays at 6,300 m. Decreased dietary intake compounded by intestinal malabsorption would account for significant reduction in exogenous energy supply, which would explain the shift to utilization of endogenous fat and protein. Significant alterations in carbohydrate, protein, and fat metabolism have been observed that correspond to the changes noted in body composition.

Carbohydrate Metabolism

The circulating level of glucose is often used as a nonspecific indicator of the state of carbohydrate metabolism. Many investigators have found lower blood glucose levels after fasting in humans exposed to moderate altitudes (below 5,000 m) for periods ranging from 2 days to 2 yr (17, 27, 32). Similar findings have been reported in studies of other mammals (9, 11). During AMREE the fasting serum glucose concentrations were lower at altitude than at sea level, but the difference was only 7 mg/100 ml (Fig. 1).

In addition to low blood sugar levels, liver glycogen content decreases in rats and mice during both acute and chronic exposures to 3,800 m (1, 36). The lower glycogen level could result from one or more of the following factors: diminished food intake, decreased intestinal absorption of glucose, increased glycogenolysis associated with an increased anaerobic glycolysis, or a generalized increase in tissue utilization of glucose. Because intravenous glucose infusion did not raise the liver glycogen content in rats at 3,800 m, Timiras et

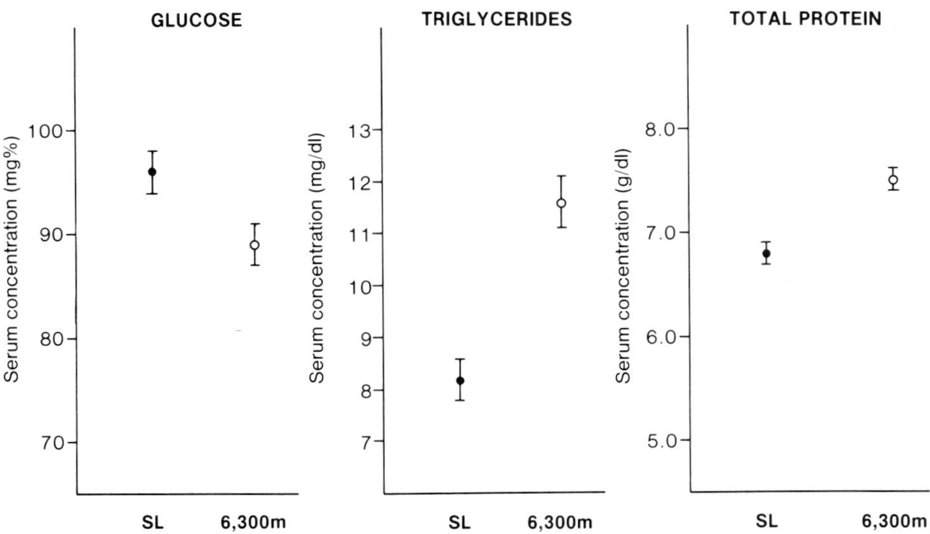

FIG. 1. Mean fasting concentrations of glucose, triglycerides, and total protein at sea level (SL; $n = 13$) and 6,300 m ($n = 13$). *Vertical lines*, ± SEM.

al. (36) suggested that neither food intake nor intestinal absorption played a major role in this alteration.

Blume and Pace (1) reported a significant decrease in glucose oxidation to carbon dioxide during the postprandial period in mice chronically exposed to an altitude of 3,800 m. That this effect was not observed during the absorptive state suggests an alteration that manifests itself when the body depends on endogenous glucose sources. In other studies Blume and Pace (3) found that the liver glycogen contained higher amounts of radiocarbon after labeled glucose injection in high-altitude rats compared to their sea-level counterparts. This suggests that glycogen synthesis is enhanced at altitude despite the lower total liver glycogen content, which implies that hepatic glycogen turnover is generally increased at altitude.

Studies have shown the rise in blood sugar levels after oral glucose-tolerance tests to be slightly more rapid at moderate altitude (32), although the overall tolerance remains near normal (4, 11). On the other hand, the AMREE glucose-tolerance data at 6,300 m demonstrated a significant change in the serum glucose response after glucose loading at 6,300 m as compared to the sea-level control response. At altitude serum glucose levels remained fairly stable after the ingestion of the glucose load. This could be explained by decreased absorption of the glucose, increased tissue removal of glucose, or a combination of both. Serum insulin concentrations increased after the loading but were generally lower than those at sea level. This could indicate that glucose was being absorbed but at a lower rate, which is consistent with the finding of intestinal malabsorption. An enhanced glucose removal, possibly due to increased tissue sensitivity to insulin, might explain the unchanged serum glucose at altitude.

Fat Metabolism

For the body to maintain or increase its level of energy expenditure at altitude despite reduced dietary intake and limited glucose storage, a shift to nonglucose energy sources seems necessary. The reports of body weight loss in association with increased circulating levels of free fatty acids and triglycerides indicate that body fat is being mobilized at altitude (18, 37). Blume and Pace (3) showed that the oxidation of ^{14}C-labeled palmitate to $^{14}CO_2$ increased in mice after a 30-day exposure to 3,800 m and that the interconversion of fatty acid radiocarbon to labeled liver glycogen was also enhanced. The increased fasting serum levels of triglycerides (Fig. 1) and the decrease in skinfold fat observed during AMREE support the view that fat is being mobilized and serves as an energy source.

Protein Metabolism

The idea that gluconeogenesis increases at altitude is also supported by the earlier findings that protein catabolism is increased while synthesis is depressed at 4,300 m, resulting in a negative nitrogen balance (5, 12). Surks et al. (35) reported that the muscle protein stores were reduced at 4,300 m, whereas the protein levels in nonmuscle tissues were increased, indicating that muscle protein mobilization is an energy source. The AMREE fasting serum levels of total protein were significantly increased at altitude (Fig. 1). This finding and the inferred decrease in muscle mass at higher elevations suggest increased protein catabolism and partly explain the physical deterioration in mountaineers at extreme altitudes.

ENDOCRINE CHANGES

The actions of a variety of hormones significantly influence utilization of carbohydrates, fats, and proteins. Earlier studies at moderate altitudes have shown elevated levels of the thyroid hormones (20, 34), catecholamines (24, 31), and glucocorticoids (14, 15). Although some of these apparently represent transient changes during the acute phase of exposure, all were studied during the chronic exposures to extreme altitudes on AMREE.

Pancreatic Hormones

Carbohydrate metabolism is largely regulated by the antagonistic actions between insulin and glucagon. During the normal diurnal fluctuations in exogenous glucose supply, these hormones, acting together, maintain the plasma glucose level within fairly narrow limits (21). The altitude-induced changes in exogenous energy supply, hepatic glycogen turnover, and tissue glucose utilization could result from alterations in the circulating levels of these hormones.

During AMREE the fasting serum levels of insulin remained very near the sea-level values at 5,400 m and were slightly elevated at 6,300 m. These findings agree with similar studies during chronic exposure to lower altitudes (10, 16). However, the lower insulin response after glucose loading at 6,300 m was in contrast to the normal response observed at moderate elevations (32). Intestinal malabsorption of glucose rather than a change in the insulin-release mechanism induced by hypoxia more likely causes the response.

Johnson et al. (17) found a significantly lower rise in blood glucose after a glucagon injection in humans at altitude than at sea level. This reflects a decrease in either hepatic glycogen content or tissue sensitivity to the hormone. On AMREE the glucagon levels were unchanged at both altitudes in the fasting state and slightly depressed after glucose loading. Thus glucagon does not appear to play a major role in the metabolic changes that were observed.

Thyroid Hormones

The principal thyroid hormones, thyroxine (T_4) and triiodothyronine (T_3), increase tissue utilization of carbohydrates and fats, increasing oxygen consumption and heat production. Changes that occur in the release mechanisms and the action of these hormones in the hypoxic state are therefore of interest. Several investigators have shown initial increases in serum T_4 and T_3 concentrations at altitude, followed by decreases toward normal with continued exposure (20, 34). Serum thyroid-stimulating hormone (TSH) levels and the pituitary release of this hormone in response to thyrotropin-releasing hormone (TRH) were normal at altitude (19, 29).

The AMREE results differ somewhat. The serum T_4 increased significantly during chronic exposure to altitude (Fig. 2), suggesting either increased secretion or decreased clearance of T_4 (23). Although the T_3 concentrations also increased at altitude, the increase was not as large as that for T_4. This resulted in an increased T_4:T_3 ratio, indicating decreased peripheral conversion of T_4 to T_3 at these extreme elevations. Increases in T_4 and T_3 levels of this magnitude normally stimulate oxidative phosphorylation, protein synthesis, and lipolysis.

The serum levels of TSH were also increased in the fasting state despite the marked elevation in T_4 levels, whereas prolactin levels were normal. Prolactin and TSH are both stimulated by TRH from the hypothalamus. Intravenous injection of TRH in fasting subjects normally elevates both TSH and prolactin levels. At altitude the response was somewhat altered; the TSH response was much higher in three of the five subjects studied, although the prolactin increase was the same as at sea level. This suggests that the pituitary feedback set point for TSH secretion was altered at altitude.

Glucocorticoid Activity

The glucocorticoids (primarily cortisol) exert metabolic effects similar to glucagon by increasing the levels of hepatic glycogenolysis and serum free fatty acid while, unlike glucagon, decreasing tissue oxidation of glucose.

Cortisol secretions rise briefly during the first few weeks at altitude (14, 15). Serum cortisol and the 17-hydroxycorticoid precursors are also increased during this period. On AMREE the diurnal cortisol levels during the chronic exposure to 6,300 m were unchanged from those noted at sea level. Thus it appears that the glucocorticoids do not have a significant metabolic effect during prolonged altitude exposure.

Catecholamines

The catecholamines epinephrine and norepinephrine have significant metabolic effects, including stimulation of glycogenolysis, glycogen turnover, and gluconeogenesis. These hormones act similar to the glucocorticoids by inhibiting insulin action and increasing glucagon release in response to hypoglycemia.

At moderate altitudes the catecholamine levels are elevated briefly during the first days of exposure but return to normal sea-level values thereafter (24,

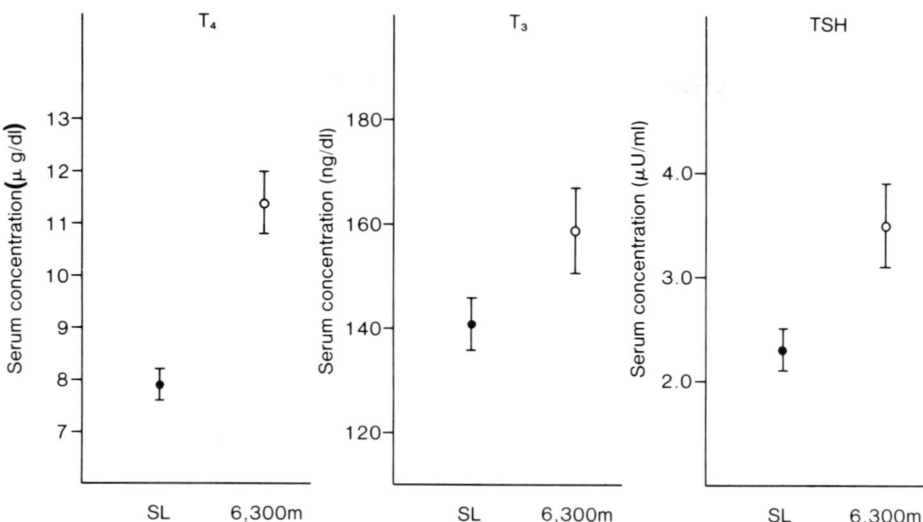

FIG. 2. Mean fasting serum concentrations of thyroxine (T_4), triiodothyronine (T_3), and thyroid-stimulating hormone (TSH) at sea level (SL; $n = 17$) and at 6,300 m ($n = 13$). *Vertical lines*, ± SEM.

31). It has been suggested that hypoxia inactivates epinephrine (26). Norepinephrine levels, however, increase during the first 2 wk at 3,800 m (25).

The AMREE studies at 6,300 m showed that norepinephrine levels were significantly higher after prolonged exposure at or above 5,400 m, whereas the epinephrine levels were normal. Thus peripheral sympathetic stimulation apparently persists during high-altitude exposure. We can only speculate on the extent to which this increase in norepinephrine influences the metabolic changes, but the gluconeogenic activity and lipid mobilization may be influenced, particularly because the glucocorticoid levels were unchanged.

Growth Hormone

It was of great interest to find that growth hormone levels remained at sea-level values in all but two AMREE members. The serum level of growth hormone was increased fivefold in the two members who experienced body weight losses of 15 kg. Although dietary intakes were not measured, the subjective reports leave little doubt that the two were in a state of severe dietary reduction. To what extent the growth hormone response is a reaction to the substantial muscle wasting in these subjects is unknown, but it might be an attempt to save protein by promoting the anabolic action of the hormone.

SUMMARY

The duration and level of high-altitude exposure cause varied effects on body metabolism. The acute effects at moderate altitudes are reflected in a transient weight loss primarily caused by the initial anorexia and respiratory water loss. As acclimatization occurs the metabolic changes are lessened in severity or disappear. As the altitude increases above 5,000 m, the alterations become more pronounced and ongoing. Apparently a combination of decreased food intake and intestinal malabsorption reduces the exogenous energy supply and forces the body to rely on its endogenous energy stores. Notable losses of body fat and protein stores may represent the utilization of these materials to meet the energy demands. To what degree the elevated levels of norepinephrine and the thyroid hormones influence these changes is not clear but must be considered in future investigations.

REFERENCES

1. BLUME, F. D., AND N. PACE. Effect of translocation to 3,800 m altitude on glycolysis in mice. *J. Appl. Physiol.* 23: 75–79, 1967.
2. BLUME, F. D., AND N. PACE. Changes in tissue distribution of glucose radiocarbon at altitude. *Federation Proc.* 28: 933–936, 1969.
3. BLUME, F. D., AND N. PACE. The utilization of ^{14}C-labeled palmitic acid, alanine and aspartic acid at high altitude. *Environ. Physiol.* 1: 30–36, 1971.
4. BRAHMACHARI, H. D., M. S. MALHOTRA, S. JOSEPH, AND U. R. KRISHNAN. Glucose tolerance at high altitude in man. *Indian J. Med. Res.* 61: 411–415, 1973.
5. CHINN, K. S. K., AND J. P. HANNON. Efficiency of food utilization at high altitude. *Federation Proc.* 28: 944–947, 1969.
6. CHINN, K. S. K., AND J. P. HANNON. Effect of diet and altitude on the body composition of rats. *J. Nutr.* 100: 732–738, 1970.
7. CONSOLAZIO, C. F., H. J. JOHNSON, AND J. H. KRZYWICKI. Body fluids, body composition and metabolic aspects of high altitude adaptation. In: *Physiological Adaptation: Desert and Mountain*, edited by M. K. Yousef, S. M. Howath, and R. W. Bullard. New York: Academic, 1972, chapt. 16.
8. CONSOLAZIO, C. F., L. O. MATOUSH, H. L. JOHNSON, AND T. A. DAWS. Protein and water balances of young adults during prolonged exposure to high altitude (4300 meters). *Am. J. Clin. Nutr.* 21: 154–161, 1968.
9. DAS, H. K., AND N. C. CHOSH. Blood sugar levels in rats exposed to varying altitude stress for different periods of time. *Aerosp. Med.* 47: 716–720, 1974.
10. DAVIDSON, M. B., AND V. S. AOKI. Fasting glucose homeostasis in rats after chronic exposure to hypoxia. *Am. J. Physiol.* 219: 378–383, 1970.
11. DRAMISE, J. G., C. M. INOUYE, B. M. CHRISTENSEN, R. D. FULTS, J. E. CANHAM, AND C. F. CONSOLAZIO. Effects of a glucose meal on human pulmonary function at 1600 m and 4300 m altitudes. *Aviat. Space Environ. Med.* 46: 365–368, 1975.
12. HANNON, J. P., C. J. KLAIN, D. M. SUDMAN, AND F. J. SULLIVAN. Nutritional aspects of high-altitude exposure in women. *Am. J. Clin. Nutr.* 29: 604–613, 1976.
13. HANNON, J. P., AND G. B. ROGERS. Body composition of mice following exposure to 4300 and 6100 meters. *Aviat. Space Environ. Med.* 46: 1232–1235, 1975.
14. HAYS, F. L., H. ARMBRUSTER, W. VETTER, AND W. BIANCA. Plasma cortisol in cattle: circadian rhythm and exposure to a simulated altitude of 5000 m. *Int. J. Biometeor.* 19: 127–135, 1975.
15. HUMPELER, E., AND F. SKRABAL. The influence of 11 days at medium altitude on the oxygen affinity, 2, 3 DPG, inorganic plasma phosphate and the plasma concentration of cortisol and aldosterone. *Pfluegers Arch. Suppl.* 377: R25, 1978.
16. JANOSKI, A. H., H. L. JOHNSON, AND S. S. SANDBAR. Carbohydrate metabolism in men at altitude (abstr.). *Federation Proc.* 28: 593, 1979.
17. JOHNSON, J. L., C. F. CONSOLAZIO, R. F. BURK, AND T. A. DAWS. Glucose-^{14}C-UL metabolism in man after abrupt altitude exposure (4300 m). *Aerosp. Med.* 45: 849–854, 1974.
18. KLAIN, G. J., AND J. P. HANNON. Effects of high altitude on lipid components of human serum. *Proc. Soc. Exp. Biol. Med.* 129: 646–649, 1968.
19. KOTCHEN, T. A., E. H. MOUGEY, R. P. HOGAN, A. E. BOYD III, L. L. PENNINGTON, AND J. W. MASON. Thyroid responses to simulated altitude. *J. Appl. Physiol.* 34: 165–168, 1973.
20. LAROCHE, G., AND C. L. JOHNSON. Simulated altitude and iodine metabolism. I. Acute effects on serum and thyroid components. *Aerosp. Med.* 38: 499–506, 1967.
21. LILJENQUIST, J. E., V. KELLER, J. L. CHIASSON, AND A. D. CHERRINGTON. Insulin and glucagon action and consequences of derangements in secretion. In: *Endocrinology*, edited by L. E. DeGroot, G. F. Cahill, W. D. Odell, L. Martini, J. T. Potts, D. H. Nelson, E. Steinberger, and A. I. Winegrad. New York: Grune & Stratton, 1979, vol. 2, chapt. 79, p. 981–996.
22. MILLEDGE, J. S. Arterial oxygen desaturation and intestinal absorption of xylose. *Br. Med. J.* 2: 557–558, 1972.
23. MORDES, J. P., F. D. BLUME, S. BOYER, M. R. ZHENG, AND L. E. BRAVERMAN. High altitude pituitary-thyroid dysfunction on Mount Everest. *New Engl. J. Med.* 308: 1135–1138, 1983.
24. MYLES, W. S., AND A. J. DRUCKER. The excretion of catecholamines in rats during acute and chronic exposure to altitude. *Can. J. Physiol. Pharmacol.* 49: 721–726, 1971.
25. PACE, N., R. L. GRISWOLD, AND B. W. GRUNBAUM. Increase in urinary norepinephrine excretion during 14 days sojourn at 3800 m elevation. *Federation Proc.* 23: 251, 1964.
26. PICÓN-REÁTEGUI, E. Effect of chronic hypoxia on the action of epinephrine in carbohydrate metabolism. *J. Appl. Physiol.* 21: 1181–1184, 1966.
27. PICON-REATEGUI, E., E. R. BUSKIRK, AND P. T. BAKER. Blood glucose in high-altitude natives and during acclimatization to altitude. *J. Appl. Physiol.* 29: 560–563, 1970.

28. PUGH, L. G. C. E. Physiological and medical aspects of the Himalayan Scientific and Mountaineering Expedition 1960–61. *Br. Med. J.* 2: 621–627, 1962.
29. RASTOGI, G. K., M. S. MALHOTRA, M. C. SRIVASTAVA, R. C. SAWHNEY, G. L. DUA, K. SRIDHARAN, R. S. HOON, AND I. SINGH. Study of the pituitary-thyroid functions at high altitude in man. *J. Clin. Endocrinol. Metab.* 43: 447–452, 1977.
30. SCHNAKENBERG, D. D., L. F. KRABILL, AND P. C. WEISER. The anorexic effect of high altitude on weight gain, nitrogen retention and body composition of rats. *J. Nutr.* 101: 787–796, 1971.
31. SINGH, I., M. S. MALHOTRA, P. K. KHANNA, R. B. NANDA, T. PURSHOTTAM, T. N. UPADHYAY, U. RADHAKRISHNAN, AND H. D. BRAHMACHARI. Changes in plasma cortisol, blood antidiuretic hormone and urinary catecholamines in high-altitude pulmonary edema. *Int. J. Biometeor.* 18: 211–221, 1974.
32. SRIVASTAVA, K. K., M. M. L. KUMRIA, S. K. GROVER, K. SRIDHARAN, AND M. S. MALHOTRA. Glucose tolerance of lowlanders during prolonged stay at high altitude and among high altitude natives. *Aviat. Space Environ. Med.* 46: 144–146, 1975.
33. SURKS, M. I. Effect of hypoxia and high altitude on thyroidal iodine metabolism in rats. *Endocrinology* 78: 307–315, 1966.
34. SURKS, M. I. Effect of thyrotropin on thyroidal iodine metabolism during hypoxia. *Am. J. Physiol.* 216: 436–439, 1969.
35. SURKS, M. I., K. S. K. CHINN, AND L. O. MATOUSH. Alterations in body composition in man after acute exposure to high altitude. *J. Appl. Physiol.* 21: 1741–1746, 1966.
36. TIMIRAS, P. S., A. A. KRUM, AND N. PACE. Body and organ weights of rats during acclimatization to an altitude of 12,470 feet. *Am. J. Physiol.* 191: 598–604, 1957.
37. WHITTEN, B. K., AND A. H. JANOSKI. Effect of high altitude and diet on lipid components of human serum. *Federation Proc.* 28: 983–986, 1969.
38. YOUNG, A. J., W. J. EVANS, A. CYMERMAN, K. B. PANDOLF, J. J. KNAPIK, AND J. T. MAHER. Sparing effect of chronic high-altitude exposure on muscle glycogen utilization. *J. Appl. Physiol.: Respirat. Environ. Exercise Physiol.* 52: 857–862, 1982.

5

Renin-Aldosterone System

JAMES S. MILLEDGE

Medical Research Council Clinical Research Centre,
Northwick Park Hospital, Harrow, England

THE EFFECT OF ALTITUDE and hypoxia on the renin-angiotensin-aldosterone system is interesting for many reasons. As a physician and mountaineer I have been puzzled for many years by the etiology of acute mountain sickness, especially high-altitude pulmonary edema (HAPE). A derangement of fluid homeostasis produced indirectly by hypoxia appeared to be one characteristic. The time lag of 6–24 h between exposure to hypoxia and the onset of symptoms suggests that hypoxia may stimulate a hormonal response which over this time course could upset fluid homeostasis enough to cause symptoms. For a time we considered the antidiuretic hormone, but, although levels are raised in cases of HAPE (4), this is more likely a result than a cause of the condition. Case reports of HAPE persistently stress the importance of strenuous exertion in the formation of this condition. We therefore decided to first study the effect of exercise typical of mountaineers on fluid balance.

In a controlled trial with a fixed diet, 6–8 h of hill walking on successive days resulted in marked sodium retention, modest water retention, and expansion of the extracellular space (including plasma volume) at the expense of the intracellular space (26). Later we showed that these changes were associated with activation of the renin-aldosterone system (11). The expansion of the extracellular space resulted in subclinical or even overt edema of the ankles, puffiness around the eyes, etc. We argued that this situation would prime the susceptible subject for the development of HAPE.

RENIN-ANGIOTENSIN-ALDOSTERONE SYSTEM

Figure 1 diagrams the renin-angiotensin-aldosterone system. Renin is released from the cells of the juxtaglomerular apparatus in the kidney in response to several stimuli, including exercise and hypoxia. The exact mechanism is still controversial, and its discussion is beyond the scope of this chapter. However, it is thought that release is in response to sympathetic

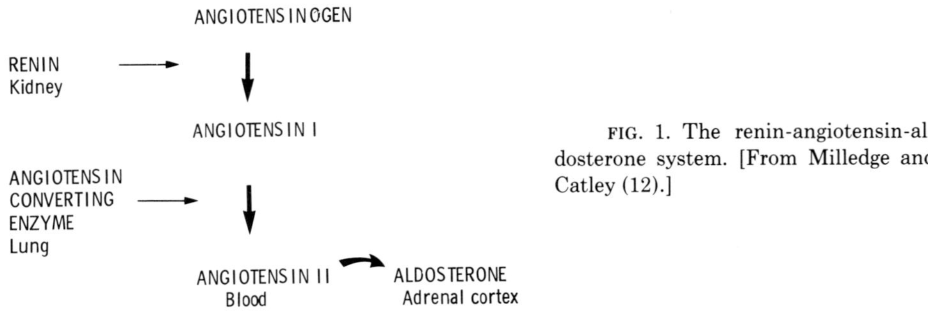

FIG. 1. The renin-angiotensin-aldosterone system. [From Milledge and Catley (12).]

stimulation and to circulating catecholamines. Hypoxia, due to local ischemia or to generalized hypoxemia, is also thought to stimulate renin release. It would not be surprising therefore if exercise at altitude resulted in very high levels of plasma renin activity (PRA). Renin itself has no biological activity but works by converting its substrate, angiotensinogen, to angiotensin I. This decapeptide is biologically inert but is converted to the vasoactive octapeptide angiotensin II by the angiotensin-converting enzyme (ACE), which is found in most endothelial tissues (especially in lung). It is a surface enzyme working on the luminal surface of the pulmonary endothelium. Conversion of angiotensin I to II is normally almost complete in one passage of blood through the lung. Angiotensin II, as well as being a potent vasoconstrictor, acts via a receptor mechanism on the adrenals to release aldosterone, which in turn acts on the renal tubules to retain sodium.

HYPOXIA

Table 1 summarizes studies on the effect of altitude hypoxia without exercise on the renin-aldosterone system. In all studies where exercise was eliminated, aldosterone levels have been reduced through a reduction in secretion rate (19). The response of PRA to altitude has been more variable. Most studies found a reduction, but in one it was unchanged and in three activity was elevated. The varying degrees of activity undertaken by subjects may have caused these differences, because exercise of some hours duration elevates PRA much more at altitude than at sea level (15). In three studies PRA and plasma aldosterone concentration (PAC) lacked concordance. While PAC was depressed, PRA was either elevated or unchanged. These studies and some of the others (7, 10) indicate a reduced response of PAC to PRA in the first few days at altitude.

How is the PAC response to PRA blunted at altitude? Leuenberger et al. (8) showed that hypoxia in dogs inhibited the activity of ACE. Perhaps this is the mechanism. We carried out a series of experiments to see *1)* if moderate hypoxia inhibited ACE activity in humans as in dogs and *2)* if such inhibition would alter the relationship between PRA and PAC. In our laboratory four subjects performed light exercise ($300\ kg \cdot m^{-1} \cdot min^{-1}$) for 2 h. During the 1st h they breathed air and in the 2nd h a hypoxic mixture (12.8% oxygen, equivalent to 4,000 m). The results are shown in Figure 2 as percents of the values at 60 min, the switch-over time from air to 12.8% oxygen. Exercise raises PAC and PRA but has no effect on ACE activity. Adding hypoxia to exercise further increases PRA but decreases PAC and ACE activity. Thus

TABLE 1. *Effect of hypoxia on aldosterone secretion and serum renin activity in man*

Exercise	Altitude Above Sea Level, m	Method for Aldosterone Study	Aldosterone Secretion	Serum or Plasma Renin Activity	Ref.
Uncontrolled					
Without diet	4,570+	Salivary Na/K	Decreased		24
	5,330	Circulating	Increased	Increased	1
	6,600+	Urine	Decreased		17
With diet	4,300	Urine Na/K	Decreased		6
Eliminated					
Without diet	4,350+	Urine	Decreased		25
With diet	3,500	Circulating and secretion	Decreased	Increased	19
	3,660	Urine	Decreased	Decreased	5
	4,279*	Urine Na/K	Decreased	Increased	22
	4,300	Circulating	Decreased	Decreased	9
	4,350	Circulating and urine	Decreased	Decreased	7
	4,760*	Circulating and urine	Decreased	No change	21
Mandatory, with diet	4,300	Circulating	Increased	Increased	9

* Equivalent heights. [Adapted from Milledge et al. (15).]

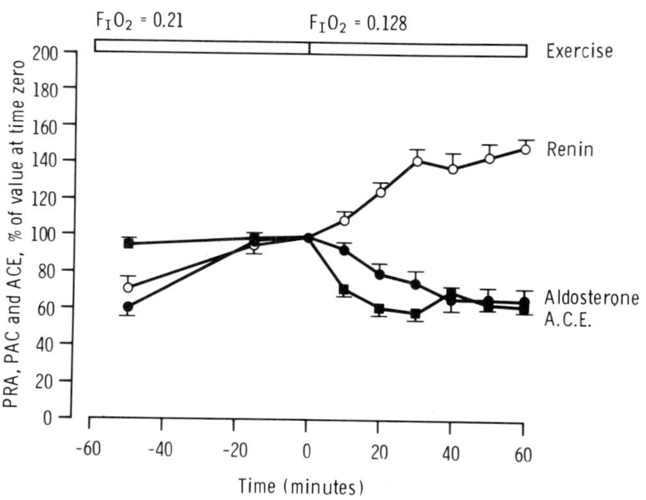

FIG. 2. Response of plasma renin activity (PRA; ○), plasma aldosterone concentration (PAC; ●), and angiotensin-converting enzyme (ACE; ■) to 120 min of exercise. First 60 min (−60 to 0 min) breathing room air [fractional concn of O_2 in dry inspired air (F_{IO_2}) of 21%] and next 60 min (0 to 60 min), a hypoxic mixture (F_{IO_2} of 12.8%). All values expressed as % response at time 0. Means ± SE for 4 subjects. [From Milledge and Catley (12).]

hypoxia reduces ACE activity and dissociates PRA and PAC. The reduction in ACE activity has a rapid onset and is complete by the 1st sampling, 10 min after the subject starts to breathe the hypoxic mixture.

The mechanism by which hypoxia reduces ACE activity is not known. Hypoxia does not directly affect the activity of the enzyme in vitro, but Stalcup et al. (20) have shown that it reduces the activity of ACE in the supernatant fluid of endothelial cell cultures. Ryan and Ryan (18) have shown that the caveolae of the endothelial cell luminal surface contain particularly high concentrations of ACE; hypoxia may produce a conformational charge of this surface, shutting off these caveolae from the passing bloodstream. As yet no experimental evidence supports this suggestion. The relationship of the serum ACE activity measured in humans to the actual capacity of the body (especially lungs) to convert angiotensin I to II must also be considered. Leuenberger et al. (8) measured this directly by infusing angiotensin I and observing its conversion under control and hypoxic conditions. The order of magnitude of the reduction in serum ACE activity in our subjects breathing low oxygen mixtures—and later during acute exposure to altitude—was similar to that reported by Leuenberger et al. (making an assumption about the degree of arterial desaturation expected). We are currently making more direct comparisons in humans by using hypoxic gas mixtures and ear oximetry. Preliminary results appear to correlate well with those of Leuenberger et al. This suggests that, although the important action of the enzyme is at the cell surface, the blood level gives an estimate of the exposure of the enzyme to the passing bloodstream—at least in the particular circumstances of acute hypoxia—and parallels the conversion capacity of the lungs.

SHORT-TERM ALTITUDE EXPOSURE WITH EXERCISE

In 1980 we mounted a field trial to study the effect of both exercise and altitude on the renin-aldosterone system (15, 27). Six male subjects on a diet of fixed electrolyte composition were studied for 13 days. The first and last 4 days were spent at low altitude (900 m), and subjects were semisedentary. For the middle 5 days they were at 3,100 m altitude on the Gornergrat in Switzerland and went hill walking for about 7 h each day. The protocol was the same as that of the previous exercise study at low altitude; five of the six subjects were the same. The results of fluid and electrolyte balances were similar to those with exercise at sea level, i.e., a marked retention of sodium, a modest water retention, and an expansion of the extracellular space, including plasma volume (resulting in a fall in hematocrit). No subject had any symptoms of acute mountain sickness, but all had subclinical leg edema as shown by a significant increase in lower leg volume by the end of the exercise-altitude period. In samples taken immediately after exercise, PRA and PAC were elevated. The elevation of PAC for at least 7 h of the day was the probable cause of the sodium and water retention.

It is interesting to compare the time course and levels of PRA and PAC in these two exercise studies at high and low altitude (Fig. 3). For PAC they are almost identical with rises on the first 2 days of exercise to almost five times control values, resulting in sodium retention of a similar magnitude (mean, 264 mmol at sea level and 202 mmol at altitude) by the end of the 5 days exercise. For PRA (*lower panel*), however, the picture is different. At sea level PRA approximately doubled on the first 2 days of exercise, whereas at altitude it rose eightfold compared with control values. Presumably the twin stimuli of unaccustomed altitude and exercise produced a powerful stimulus to renin output. However, these high levels of PRA do not cause similar high levels of PAC. Replotted as percentages of their control values these results show that the PAC response to PRA at altitude is blunted (Fig. 4). The slope of the response at sea level is 3 and at altitude 0.56.

We also measured ACE activity in these studies (Fig. 5). Exercise alone has no effect, but even the modest altitude of our base station for the latter study (900 m) measurably affected ACE activity. The altitude of 3,100 m, still modest by Himalayan standards, markedly lowers ACE activity by a further 23%. We suggest that this reduction in ACE activity is the mechanism for the immediate reduction in the PAC response to PRA. This blunting of response would be beneficial during extremely elevated PRA because high levels of angiotensin II would increase blood pressure in both systemic and pulmonary circulation and because high levels of aldosterone would result in even more marked sodium retention. These changes would tend to place the subject at an increased risk of edema formation, including pulmonary and possibly cerebral edema. Hence we speculated that a constitutional failure to lower

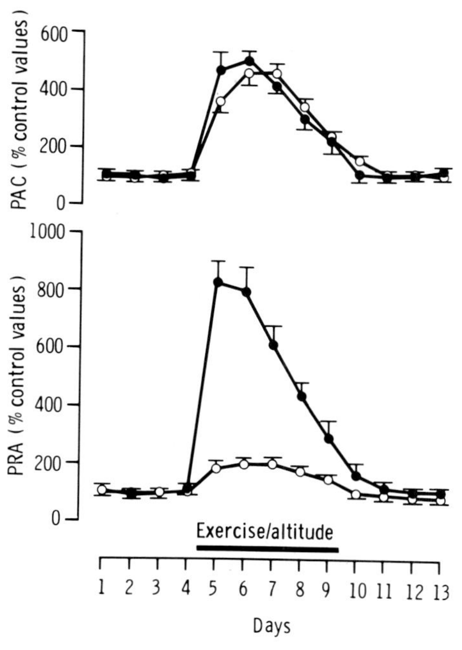

FIG. 3. Comparison of changes in PRA (*lower panel*) and PAC (*upper panel*) during exercise at 3,100 m (●) and with exercise at low altitude (○). Means ± SE. [From Frayser et al. (1).]

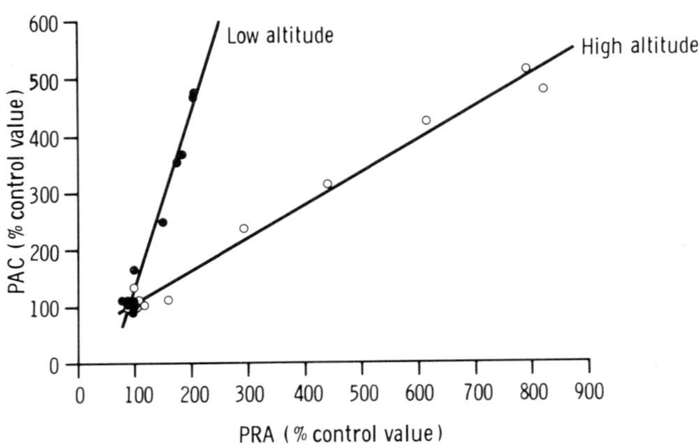

FIG. 4. Response of PAC to increased PRA. Mean results for single days from 6 subjects in the high-altitude study (○) and 5 subjects in the low-altitude study (●). [From Frayser et al. (1).]

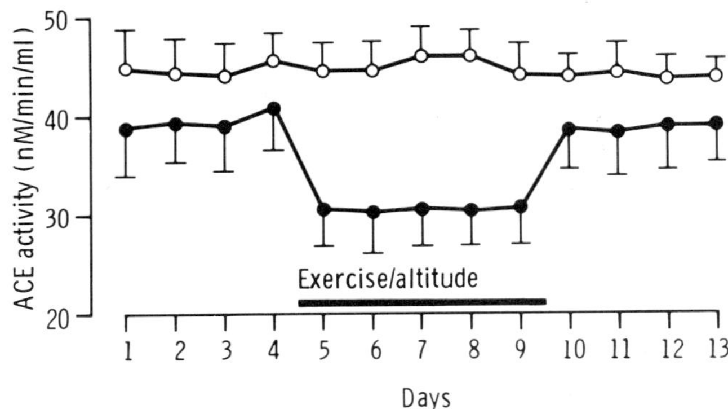

FIG. 5. Activity of ACE during exercise study at sea level (○) and at 3,100 m (●). In both studies during first and last 4 days subjects were semisedentary. In the altitude study these periods were spent at 900 m. Means ± SE. [From Frayser et al. (1).]

ACE activity with hypoxia might increase a subject's susceptibility to acute mountain sickness.

LONG-TERM ALTITUDE EXPOSURE AT REST

In 1981 we were able to study the effect of more prolonged altitude exposure on the renin-aldosterone system on an expedition to Mount Kongur in western China. Blood samples were taken at Base Camp (4,500 m) at intervals during 7 wk. Subjects throughout this period were at or above this altitude and were actively involved in climbing or hill walking on most days. However, when "at rest" samples were taken (on "rest days"), subjects engaged in normal activities around the camp. Figure 6 shows the results of PRA and PAC during this period, together with control values. Initially when subjects had been at Base Camp for 5 days and above 3,500 m for the previous 5 days,

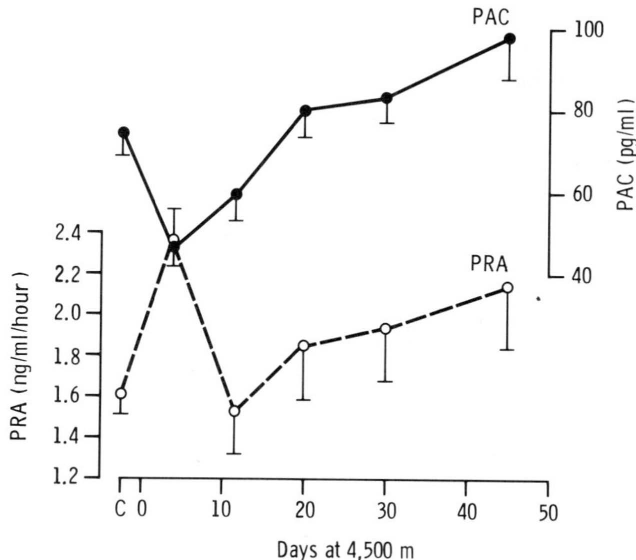

FIG. 6. Time course of PRA (○) and PAC (●) at or above 4,500 m. C, sea-level control. Means ± SE for 10 subjects. [From Milledge et al. (14).]

PAC was reduced as all previous work would predict, whereas PRA was elevated as some previous studies have found. Thereafter PRA drops back to control values and remains at about that level (the modest elevation, especially at the last sampling time, is contributed mainly by the 4 climbers who had been very active on the day before sampling). During the stay at altitude PAC recovers, reaching control values between 12 and 20 days.

The PAC:PRA ratio gives some measure of the adrenal response to PRA. In Figure 7 this ratio is plotted against time (*upper curve*). It drops from control values early in altitude exposure and recovers during the stay at altitude. Also plotted in Figure 7 are the results of ACE activity. Again the initial fall that we had expected was not maintained, but recovery to control values took place in 12–20 days followed by an overshoot so that, at 7 wk, ACE was 123% of control values. Some early work on ACE activity in mice (16) had shown this chronic effect of hypoxia at 9 and 11 days, having missed the depressant effect of acute hypoxia. Also four of our subjects who were climbers frequently exposed to extreme altitude had significantly greater depression of their ACE activity during the initial 12 days at altitude than did the other six subjects. The time course for the PAC:PRA ratio is similar to that of ACE activity, further evidence that ACE activity controls the PAC response to PRA.

LONG-TERM ALTITUDE EXPOSURE WITH EXERCISE

On Mount Kongur we took samples from subjects immediately after arrival at Base Camp from higher or lower camps. These were taken early in

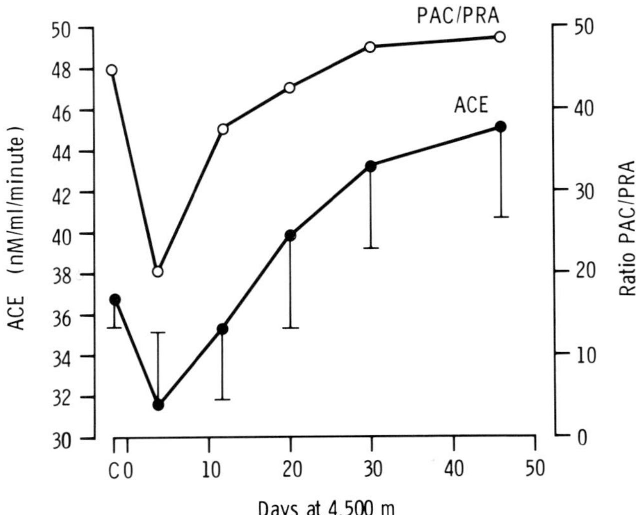

FIG. 7. Time course of change in PAC:PRA ratio (○; mean values) and of ACE activity (●; means ± SE) at or above 4,500 m for 10 subjects. C, sea-level control. [From Milledge et al. (14).]

the expedition when ACE activity was still depressed. Like those of the Gornergrat study, these results showed blunting of the PAC response to PRA.

Later in 1981 there was a further opportunity to collect samples both at rest and after exercise in subjects exposed to prolonged altitude at even higher elevations. This was on the 1981 American Medical Research Expedition to Everest. In resting subjects at sea level and at 6,300 m after 2–4 wk at or above that altitude PRA, PAC, and ACE activity were measured. By this time all of these hormones had returned to within 10% of sea-level values.

In seven subjects samples were taken immediately after arrival at Base Camp from camps above or below. These exercise samples showed elevated PRA and PAC (Fig. 8). The PAC response to PRA is even more blunted at this higher altitude and may have flattened out. At this time ACE activity had returned to the subjects' control sea-level values. Therefore we cannot ascribe this blunted response to reduced ACE activity.

We suggest that after several weeks at altitude downregulation of angiotensin II receptors causes the blunted PAC response to PRA. With continued or frequent high levels of angiotensin II, the density of angiotensin II receptors on the adrenals is reduced and hence the response of PAC to angiotensin II is reduced. The response should also become nonlinear, with a near-normal response at low levels of angiotensin II, decreasing to a flat response as all the receptors become occupied. As noted, there is a hint that this happened in our Everest data, because the PAC:PRA ratio is normal at rest, whereas exercise points show marked blunting and a rather flat response. This downregulation is a common phenomenon of receptor systems and occurs in insulin receptors on adipocytes exposed to high levels of insulin (3) and in adrenergic receptors on heart muscle in rats exposed to chronic hypoxia (23). The time required for this mechanism is about 1–2 wk.

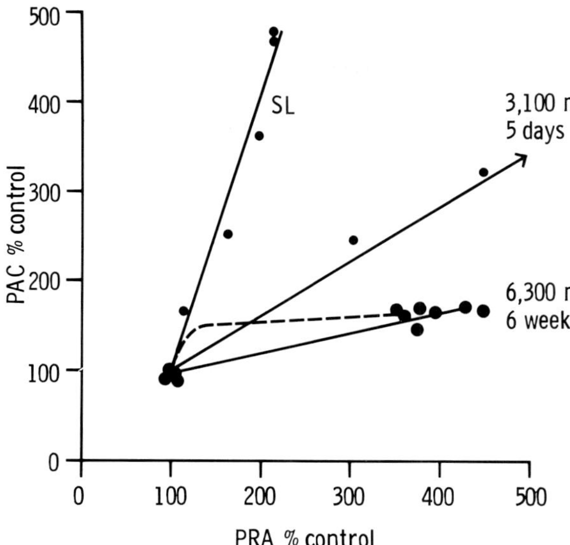

FIG. 8. Response of PAC to PRA as in Fig. 4 but with additional data from 7 subjects at or above 6,300 m for 2–4 wk. Each *large point* represents 1 subject: *points* to *left*, at rest; to *right*, with exercise. The response may be linear or nonlinear (*dashed line*). [From Milledge et al. (13).]

Another possible mechanism is the induction of enzymes that degrade angiotensin II, e.g., aminopeptidase A. This would be analogous to the induction of *O*-methyltransferase in the heart muscle of chronically hypoxic mice (9); this enzyme breaks down catecholamines, which provides another mechanism for the blunted sympathetic response of chronic hypoxia. The two mechanisms, downregulation and enzyme induction, are of course not mutually exclusive. Their time courses are similar, but with downregulation one would expect high levels of angiotensin II, whereas if enzyme induction was the major mechanism, angiotensin II concentration would be as low as that of aldosterone.

CONCLUSION

If the renin-aldosterone system is stimulated for several hours by exercise at low or high altitude a significant amount of sodium is retained. In turn the extracellular space expands, partly at the expense of the intracellular space, which provides one factor for the initiation of edema (Fig. 9). This factor may be important in the genesis of acute mountain sickness and especially HAPE.

Hypoxia is synergistic to exercise in stimulating renin output. The body normally defends itself against the consequences of high levels of renin, causing equivalent high levels of aldosterone (and angiotensin II), by rapidly reducing ACE activity with the onset of hypoxia. Failure to reduce ACE activity may also be a factor in acute mountain sickness and HAPE. Experienced climbers had a greater reduction in ACE in the first few days at altitude than less experienced subjects.

Downregulation of angiotensin II receptors may provide the long-term

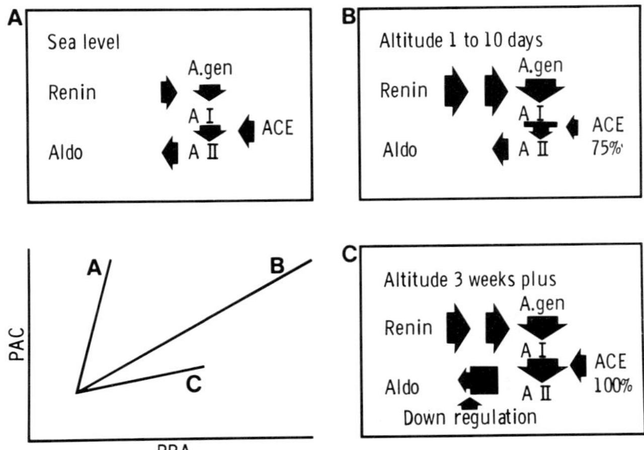

FIG. 9. Activation of renin-aldosterone system by exercise at sea level and during acute and chronic altitude exposure. Size of *arrow* indicates changes in relative activities of various components of system. Resulting changes in PAC response to PRA are indicated in graph, *bottom left.*

defense against high levels of aldosterone, although it would not provide defense against high levels of angiotensin II during exercise at altitude and may account for the occasional high blood pressure found in subjects after exercise at altitude. It may also be a factor in the strokes occasionally reported in otherwise fit young men at altitude. Alternatively enzymes that degrade angiotensin II may be induced.

This hypothesis is diagramed in Figure 9 and, although presented tentatively, best explains our findings. The figure also points to further studies that would support or refute these possible factors.

REFERENCES

1. FRAYSER, R., I. D. RENNIE, G. W. GRAY, AND C. S. HOUSTON. Hormonal and electrolyte response to exposure to 17,500 ft. *J. Appl. Physiol.* 38: 636–642, 1975.
2. FRIEDLAND, J., AND E. SILVERSTEIN. A sensitive fluorimetric assay for serum angtiotensin-converting enzyme. *Am. J. Clin. Pathol.* 66: 416–424, 1976.
3. GAVIN, J. R., J. R. ROTH, D. M. NEVILLE, P. DE MEYTS, AND D. N. BUELL. Insulin-dependent regulation of insulin receptor concentrations: a direct demonstration in cell culture. *Proc. Natl. Acad. Sci. USA* 71: 84–88, 1974.
4. HACKET, P. H., M. L. FORSLING, J. S. MILLEDGE, AND D. RENNIE. Release of vasopressin in man at altitude. *Horm. Metab. Res.* 10: 571, 1978.
5. HOGAN, R. P. III, T. A. KOTCHEN, A. E. BOYD III, AND L. H. HARTLEY. Effect of altitude on renin-aldosterone system and metabolism of water and electrolytes. *J. Appl. Physiol.* 35: 385–390, 1973.
6. JANOSKI, A. H., B. K. WHITTEN, J. L. SHIELDS, AND J. P. HANNON. Electrolyte patterns and regulation in man during acute exposure to high altitude. *Federation Proc.* 28: 1185–1189, 1969.
7. KEYNES, R. J., G. W. SMITH, J. D. H. SLATER, M. M. BROWN, S. E. BROWN, N. N. PAYNE, T. P. JOWETT, AND C. C. MONGE. Renin and aldosterone at high altitude in man. *J. Endocrinol.* 92: 131–140, 1982.
8. LEUENBERGER, P. J., S. A. STALCUP, R. B. MELLINS, L. M. GREENBAUM, AND G. M. TURINO. Decrease in angiotensin I conversion by acute hypoxia in dogs. *Proc. Soc. Exp. Biol. Med.* 158: 586–589, 1978.
9. MAHER, J. T., J. C. DENNISTON, D. L. WOLFE, AND A. CYMERMAN. Mechanism of the attenuated cardiac response to β-adrenergic stimulation in chronic hypoxia. *J. Appl. Physiol.: Respirat. Environ. Exercise Physiol.* 44: 647–651, 1978.
10. MAHER, J. T., L. G. JONES, L. H. HARTLEY, G. H. WILLIAMS, AND L. I. ROSE. Aldosterone dynamics during graded exercise at sea level and high altitude. *J. Appl. Physiol.* 39: 18–22, 1975.
11. MILLEDGE, J. S., E. I. BRYSON, D. M. CATLEY, R. HESP, N. LUFF., B. D. MINTY, M. W. J. OLDER, N. N. PAYNE, M. P. WARD, AND W. R. WITHEY. Sodium balance, fluid homeostasis and the renin-aldosterone system during the prolonged exercise of hill walking. *Clin. Sci.* 62: 595–604, 1982.
12. MILLEDGE, J. S., AND D. M. CATLEY. Renin, aldosterone, and converting enzyme during exercise and acute

hypoxia in humans. *J. Appl. Physiol.: Respirat. Environ. Exercise Physiol.* 52: 320–323, 1982.

13. MILLEDGE, J. S., D. M. CATLEY, F. D. BLUME, AND J. B. WEST. Renin, angiotensin-converting enzyme, and aldosterone in humans on Mount Everest. *J. Appl. Physiol.: Respirat. Environ. Exercise Physiol.* 55: 1109–1112, 1983.
14. MILLEDGE, J. S., D. M. CATLEY, M. P. WARD, E. S. WILLIAMS, AND C. R. A. CLARKE. Renin-aldosterone and angiotensin-converting enzyme during prolonged altitude exposure. *J. Appl. Physiol.: Respirat. Environ. Exercise Physiol.* 55: 699–702, 1983.
15. MILLEDGE, J. S., D. M. CATLEY, E. S. WILLIAMS, W. R. WITHEY, AND B. D. MINTY. The effect of prolonged exercise at altitude on the renin aldosterone system. *J. Appl. Physiol.: Respirat. Environ. Exercise Physiol.* 55: 413–418, 1983.
16. MOLTENI, A., R. M. ZAKHEIM, K. B. MALLIS, AND L. MATTIOLI. The effect of chronic alveolar hypoxia in lung and serum angiotensin I converting enzyme activity. *Proc. Soc. Exp. Biol. Med.* 147: 263–265, 1974.
17. PINES, A., J. D. H. SLATER, AND T. P. JOWETT. The kidney and aldosterone in acclimatization at altitude. *Br. J. Dis. Chest* 71: 203–297, 1977.
18. RYAN, J. W., AND U. S. RYAN. Pulmonary endothelial cells. *Federation Proc.* 36: 2683–2691, 1977.
19. SLATER, J. D. H., R. E. TUFFLEY, E. S. WILLIAMS, C. H. BERESFORD, P. H. SÖNKSEN, R. H. T. EDWARDS, R. P. EKINS, AND M. McLAUGHLIN. Control of aldosterone secretion during acclimatization to hypoxia in man. *Clin. Sci.* 37: 327–341, 1969.
20. STALCUP, S. A., J. S. LIPSET, J. M. WOUN, P. LEUENBERGER, AND R. B. MELLINS. Inhibition of angiotensin converting enzyme activity in cultured endothelial cells by hypoxia. *J. Clin. Invest.* 63: 966–976, 1979.
21. SUTTON, J. R., G. W. VIOL, G. W. GRAY, M. McFADDEN, AND P. M. KEANE. Renin, aldosterone, electrolyte, and cortisol responses to hypoxic decompression. *J. Appl. Physiol.: Respirat. Environ. Exercise Physiol.* 43: 421–424, 1977.
22. TUFFLEY, R. E., D. RUBENSTEIN, J. D. H. SLATER, AND E. S. WILLIAMS. Serum renin activity during exposure to hypoxia. *J. Endocrinol.* 48: 497–510, 1970.
23. VOELKEL, N. F., L. HEGSTRAND, J. T. REEVES, I. F. McMURTY, AND P. B. MOLINOFF. Effects of hypoxia on density of β-adrenergic receptors. *J. Appl. Physiol.: Respirat. Environ. Exercise Physiol.* 50: 363–366, 1981.
24. WILLIAMS, E. S. Salivary electrolyte composition at high altitude. *Clin. Sci.* 21: 37–42, 1961.
25. WILLIAMS, E. S. Electrolyte regulation during the adaptation of humans to life at high altitude. *Proc. R. Soc. London Ser. B* 165: 266–280, 1966.
26. WILLIAMS, E. S., M. P. WARD, J. S. MILLEDGE, W. R. WITHEY, M. W. J. OLDER, AND M. L. FORSLING. Effect of the exercise of seven consecutive days hill-walking on fluid homeostatis. *Clin. Sci.* 56: 305–316, 1979.
27. WITHEY, W. R., J. S. MILLEDGE, E. S. WILLIAMS, B. D. MINTY, E. I. BRYSON, N. P. LUFF, M. W. J. OLDER, AND J. M. BEELEY. Fluid and electrolyte homeostasis during prolonged exercise at altitude. *J. Appl. Physiol.: Respirat. Environ. Exercise Physiol.* 55: 2–417, 1983.

6

Red Cell Function at Extreme Altitude

ROBERT M. WINSLOW

Division of Host Factors, Center for Infectious Diseases, Centers for Disease Control, Public Health Services, United States Department of Health and Human Services, Atlanta, Georgia

THE RELATIONSHIP BETWEEN THE AMOUNT of oxygen bound to hemoglobin (saturation) and the oxygen physically dissolved in solution [oxygen tension (Po_2)] under equilibrium conditions is the familiar oxygen equilibrium curve (OEC) of hemoglobin. Its position is often denoted by the value P_{50}, the Po_2 at which saturation is 50%. If oxygen affinity increases, the OEC shifts left and P_{50} is reduced. If oxygen affinity decreases, the OEC shifts right and P_{50} is increased. The principal effectors of hemoglobin function [H^+, 2,3-diphosphoglycerate (2,3-DPG), and carbon dioxide] all shift the OEC to the right.

Early in this century a fundamental difference between the oxygenation properties of myoglobin and hemoglobin was recognized (11): the former is hyperbolic, and the latter is sigmoid (Fig. 1). The reason for the sigmoid shape of the hemoglobin curve was discovered much later, although it is this peculiar shape that allows for the efficient transport of oxygen. Because the binding of each oxygen molecule to hemoglobin apparently facilitated the binding of subsequent molecules, early biochemists used the term "cooperativity" to describe the sigmoid shape.

Over the past 50–75 years many scientists studying the regulatory role of hemoglobin in the oxygen transport process presumed that shifts in the position and shape of the OEC adapt humans to hypoxia. However, heme-containing proteins are ubiquitous in the plant and animal kingdoms, and almost certainly hemoglobin evolved as an oxygen carrier before humans emerged from the evolutionary competition. There is no reason to believe that the behavior of the OEC at high altitude is good or bad or a uniquely human property. To be objective one must observe as accurately as possible the in vivo properties of hemoglobin, then try to place these properties in the overall scheme of oxygen transport. Only then can the relative importance of shifts to the left or right of the OEC be assessed.

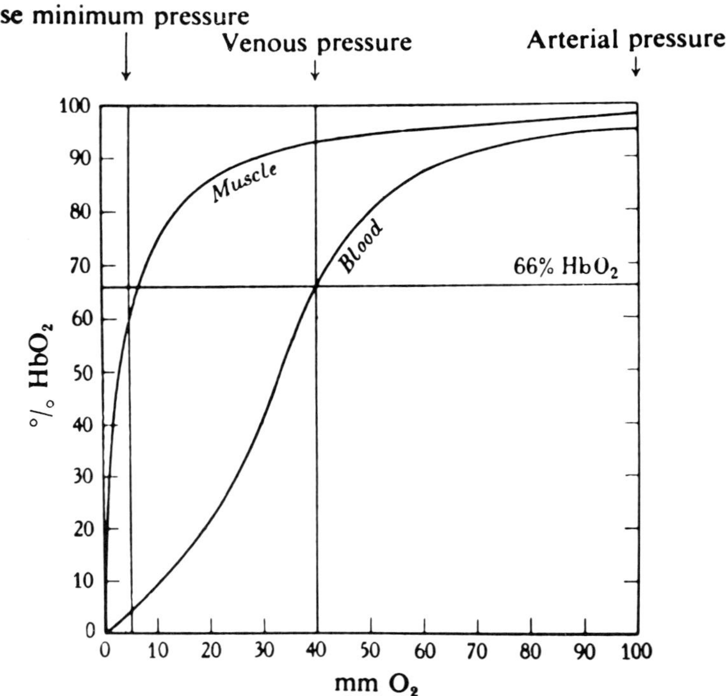

FIG. 1. Oxygen equilibrium curves of hemoglobin and myoglobin.

Because the shape of the hemoglobin curve is the key to understanding its regulation, a brief description of it is given. Recent data collected at high altitude are then reviewed.

HEMOGLOBIN STRUCTURE AND FUNCTION

Hemoglobin is a protein molecule made up of four polypeptide subunits, each of which is linked to an iron-containing tetrapyrrole prosthetic group, heme. Because oxygen binds reversibly to the iron atom, each hemoglobin molecule can bind four molecules of oxygen. The protein subunits have a molecular weight of approximately 16,100 and are structurally very similar to myoglobin.

In the normal adult human approximately 97.5% of the red cell protein is hemoglobin A, which contains two α- and two β-chains in a tetrameric structure. As X-ray crystallographic studies (3) show, hemoglobin can take up one of two conformations, the difference between them being the strength of the intersubunit chemical bonds. Although certain details of the mechanism of oxygen binding to hemoglobin are unclear, the most useful current model is based on these two conformations. In one the subunits are maximally bonded and the structure is "tense," or in the T conformation. In the other conformation some of the intersubunit bonds are ruptured and the structure is "relaxed," or in the R conformation. Molecular oxygen more easily enters the region of the heme in the latter structure.

At all stages of oxygenation, an equilibrium exists between the T and R conformations. However, the position of the equilibrium is strongly influenced by the number of oxygen molecules bound. Nearly all of the deoxygenated hemoglobin is in the T conformation, whereas nearly all of the oxyhemoglobin is in the R conformation. The OEC has a sigmoid shape because the molecule in the R state has a much higher affinity for oxygen than that in the T state. To visualize this interaction, consider the successive binding of oxygen to hemoglobin. Deoxyhemoglobin is in the T state. As the first oxygen molecule is bound, the hemoglobin molecule is forced open a bit, rupturing one or more of the intersubunit bonds, and some of the molecules change to the R conformation. These have a greatly increased oxygen affinity, the second oxygen molecule is bound more easily, and so on. On the average, most of the molecules switch from T to R between the second and third oxygen, causing the OEC to be steepest in its central portion. As shown in Figure 1, this is extremely important because small changes in Po_2 in this steep portion lead to large changes in oxygen saturation.

The position of the OEC under physiological conditions is regulated almost entirely by three effectors: H^+, 2,3-DPG, and carbon dioxide. Each of these molecules stabilizes the T conformation. Hydrogen ions contribute to the strength of the intersubunit bonds (Bohr effect); 2,3-DPG binds between β-chains to increase the number of intersubunit bonds, as does, to a smaller extent, carbon dioxide. Increasing the number of intersubunit bonds decreases oxygen affinity and increases the Bohr effect. Thus, depending on the pH, increases in 2,3-DPG or carbon dioxide tension (Pco_2) shift the OEC to the right. Obviously the interactions among these effector molecules are complex.

BLOOD OXYGEN EQUILIBRIUM CURVE AT ALTITUDE

The details of the molecular mechanisms of hemoglobin-oxygen interactions have been clarified only in the past 15–20 years. However, Barcroft et al. (1) recognized as early as 1922 that the position of the OEC could be important in acclimatization to high altitude. They suggested that increased oxygen affinity (left shift of OEC) might be beneficial, but early experimental studies (12) demonstrated a right shift in both sojourners and natives. Whereas Barcroft supposed that a left shift would facilitate loading of oxygen in the lungs, subsequent workers suggested that a right shift would increase the ability of the blood to unload oxygen in the tissues. Lenfant and Sullivan (14) provided the modern view that the importance of the position of the OEC greatly depends on the subject's ventilation and the altitude or degree of hypoxia.

The best explanation for the apparent contradictions in experimental data is that the older methods for measuring hemoglobin oxygen saturation and Po_2 were not suited to field studies. Until the late 1950s few used the

Clark electrode; before that (and, indeed, until the late 1960s) the tedious manometric method of Van Slyke was the accepted standard (7). Moreover the profound effect of 2,3-DPG on hemoglobin function was not recognized until 1968 (2, 5); only then was it realized that blood samples must be handled very carefully to ensure that 2,3-DPG concentration would be preserved.

Rossi-Bernardi and co-workers (17) introduced a new method to measure the whole blood OEC under strict control of 2,3-DPG, pH, and P_{CO_2}. They used a Clark electrode to measure P_{O_2} in a closed sample of blood that was first fully deoxygenated by equilibration with known mixtures of carbon dioxide and nitrogen, then gradually oxygenated by adding hydrogen peroxide at a controlled rate. By simultaneously monitoring P_{CO_2}, pH was held constant by titrating the released Bohr protons and adding sodium hydroxide under feedback control. This method was fully automated so that an experiment could be controlled and the results calculated by a small computer (28). The resulting OEC data were compared in detail with the best available equilibrium data (30). It was established that very accurate measurements could be made with fresh blood samples within ½ h of collection and that the resulting OECs were extensive enough to allow detailed analyses of the intracellular effects of H^+ and 2,3-DPG (20).

The first application of the method at high altitude was in Morococha, Peru, altitude 4,520 m, the site of early observations by Barcroft and a series of succeeding high-altitude expeditions. The laboratory in Morococha is accessible and modern, and electricity is available most of the time. Thus it was possible to set up the computerized apparatus and measure OECs with fresh blood with the same techniques used at sea level. The results reconciled previous opposing views: the OEC in Morococha is right shifted under standard conditions (pH 7.4; P_{CO_2} 40 mmHg) due to an increase in 2,3-DPG, but this is offset in vivo by a left shift caused by respiratory alkalosis. Thus the distribution of P_{50} values at Morococha and sea level are not significantly different (see Fig. 2).

The behavior of the OEC under extremely hypoxic conditions remained to be measured. Morococha is close to the maximum altitude for permanent human habitation. Consequently studies at greater altitudes must be done with sojourners. Comparison of natives and sojourners may not be entirely justified because of the changes that occur with long-term exposure to altitude (12). Also natives and sojourners have quite different motivations for trying to adapt to hypoxia: the former for sustained work at low levels and the latter for short-term high-level performance such as climbing or other athletic activities.

The 1981 American Medical Research Expedition to Everest provided the opportunity to study the behavior of the OEC at the ultimate altitude. However, in this case the measurements could not be made with the apparatus used in Morococha; two formidable obstacles had to be overcome. First, the regulation of the normal OEC by H^+, 2,3-DPG, and carbon dioxide had to be understood well enough so that the minimal amount of data to be collected on the mountain could be defined. Second, methods had to be developed that

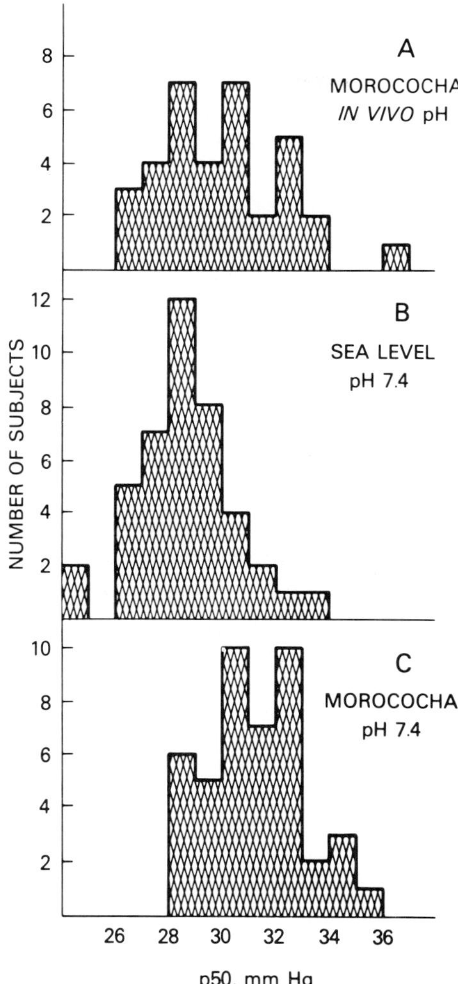

FIG. 2. Distribution of O_2 affinity at Morococha and at sea level. When corrected to pH 7.4 (*C*), values are higher at Morococha than at sea level (mean 31.2 and 29.4 mmHg, respectively). Corrected to subject's plasma pH (*A*), in vivo P_{50} falls in sea-level range in all but 1 case.

could be used under primitive conditions but would still give accurate results. A detailed analysis of the entire OEC over the full range of physiological conditions that might be encountered accomplished the first (29). It allowed the OEC to be predicted from known pH, P_{CO_2}, and 2,3-DPG. The second was accomplished by developing a new tonometry system for small blood samples that required only simple equipment (18) and a new method for accurately measuring hemoglobin saturation that would be easy to use in the field and would not be sensitive to variations of temperature or power (19).

AMERICAN MEDICAL RESEARCH EXPEDITION TO EVEREST

Studies were done in three principal locations: San Diego (May, 1981), Base Camp (September, 1981), and Camp II (middle and late October, 1981).

The locations and elevations of these various sites are diagramed in Figure 3. Care was taken to use the same techniques, instruments, and reagent stocks in each location. Instruments, packed and transported in special lightweight containers, were carried at various times by ship, truck, plane, backpack, and yak. West (24) has described the laboratory huts. The hematology laboratory at Camp II is shown in Figure 4.

Venous blood samples were taken without stasis into a mixture of heparin and sodium fluoride to inhibit glycolysis and were immediately placed on ice. Hematocrit was measured with a battery-powered Compur M1100 minicentrifuge with a radius of centrifugation of 3.46 cm and a speed of 11,500 rpm. Samples were subjected to repeated centrifugation cycles (3 min 30 s each) until the reading was constant. Total hemoglobin concentration was measured either with Drabkin's reagent in a Perkin-Elmer model 35 spectrophotometer or with a battery-powered Compur M1000 miniphotometer. The methods gave the same results. Red blood cell counts were made with the latter instrument according to the instructions of the manufacturer.

At Base Camp, 2,3-DPG was measured in neutralized acid extracts of whole blood with reagent kits from Boehringer Mannheim and the Perkin-Elmer spectrophotometer. Control experiments showed that the extracts were stable for up to 5 days. Extracts made at Camp II were transported by hand in a Dewar flask containing an ice-water slurry through the Khumbu Icefall to Base Camp, usually the day after preparation; the delay was never longer than 2 days.

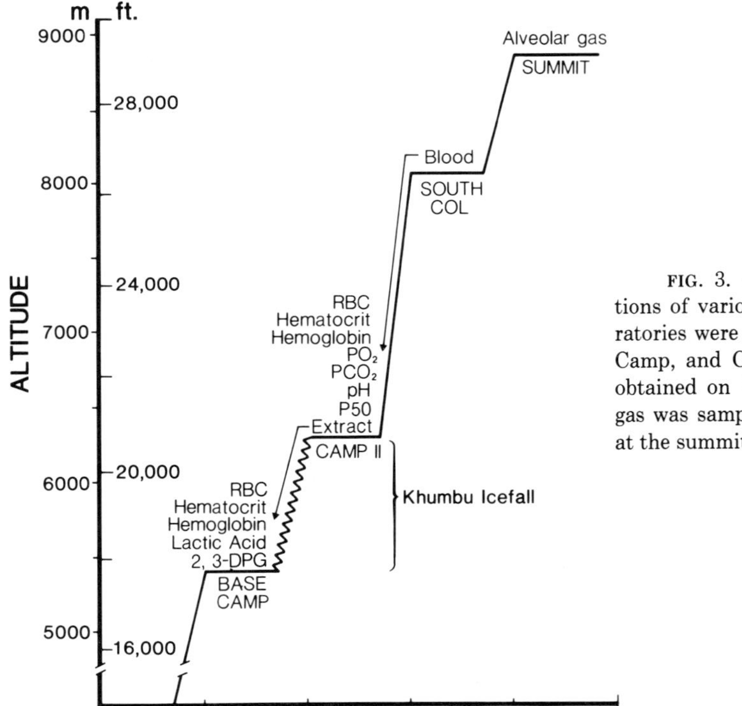

FIG. 3. Altitudes and locations of various activities. Laboratories were at San Diego, Base Camp, and Camp II. Blood was obtained on South Col; alveolar gas was sampled on the Col and at the summit.

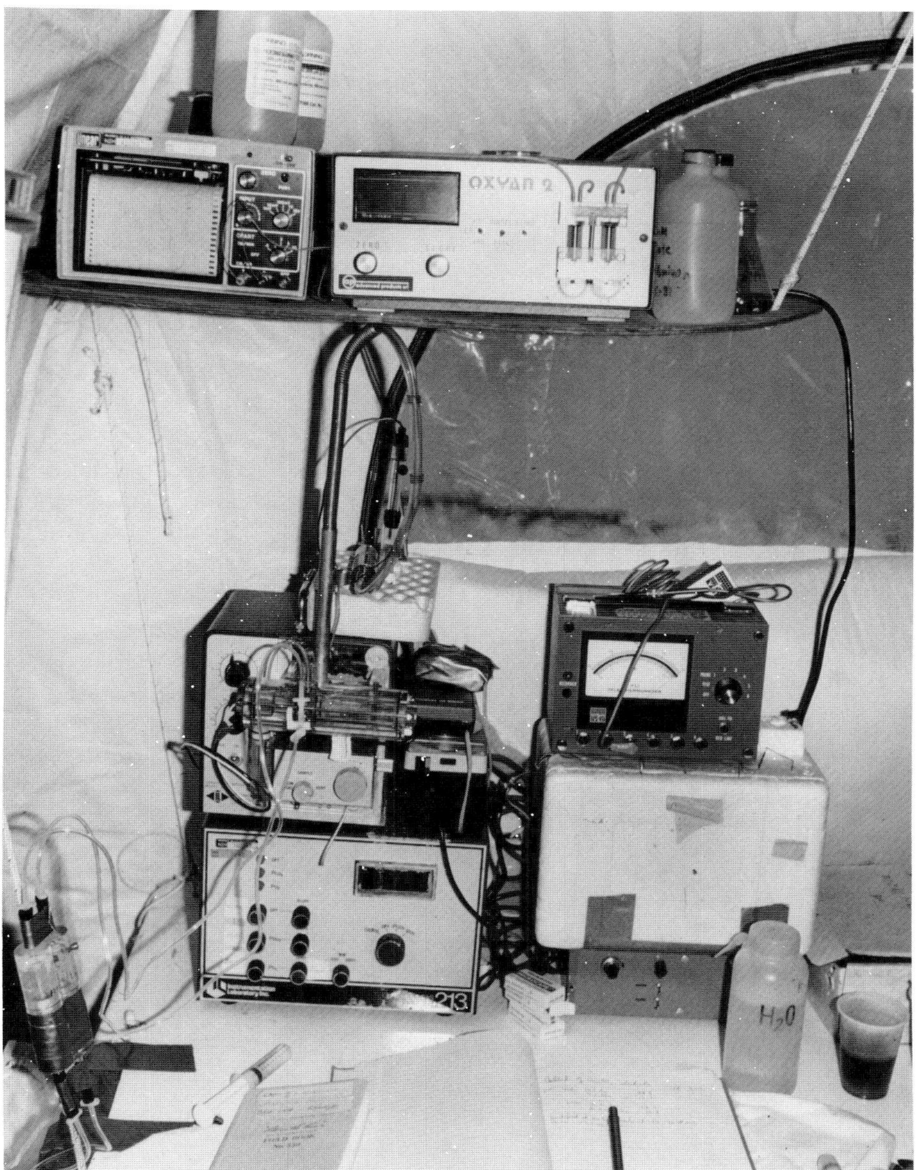

FIG. 4. Hematology section of Camp II Laboratory.

Blood gas and acid-base measurements were made at sea level and Camp II with an Instrumentation Laboratory model 213 blood-gas apparatus, electrodes, and buffers.

Oxygenated fresh blood (250 μl) was injected into a tonometry flask filled with gas mixtures of known oxygen and carbon dioxide composition to approximate anticipated P_{50} and arterial P_{CO_2} (Pa_{CO_2}). The samples were equilibrated in a tonometer at 37°C for 20 min, and saturation was then measured by the procedure detailed below. Then P_{50} was calculated with the Hill equation (10).

Measurements of oxygen saturation, made with an instrument specially constructed for the expedition, are described in detail elsewhere (19). It consists of a closed, 1-ml cuvette containing 0.01% sodium borate and 0.1% Sterox (pH 10.1) in equilibrium with air. An oxygen electrode monitored the PO_2 of the contents. When a sample of blood (10 μl) is introduced into the cuvette, the PO_2 drops in proportion to the amount of deoxyhemoglobin present because dilution, lysis, and high pH increase the hemoglobin-oxygen affinity. Then an excess of potassium ferricyanide is added to completely oxidize the hemoglobin and release all bound oxygen; the rise in observed PO_2 is proportional to total hemoglobin. The ratio (total rise − initial drop)/total rise is fractional saturation (see Fig. 5). The calculations, assumptions, and controls are discussed elsewhere (31). This instrument, weighing only a few pounds, requires very little power and can be operated with a 12-V battery if necessary. Blood P_{50} was calculated according to the general method described previously by Samaja et al. (18).

Hematocrit, Hemoglobin, and Mean Corpuscular Hemoglobin Concentration

As expected, hematocrit and hemoglobin concentration increased over the period of exposure to altitude (Table 1). In general, if a person's hematocrit was in the lower part of the normal range at sea level, it remained in the lower range as altitude increased. In one exceptional individual, hematocrit increased from 42% at sea level to 64% at Camp II, and he had symptoms of bronchospasm. No correlation between absolute hematocrit (or relative change) and performance in climbing or success in acclimatization was noted within the rest of the group.

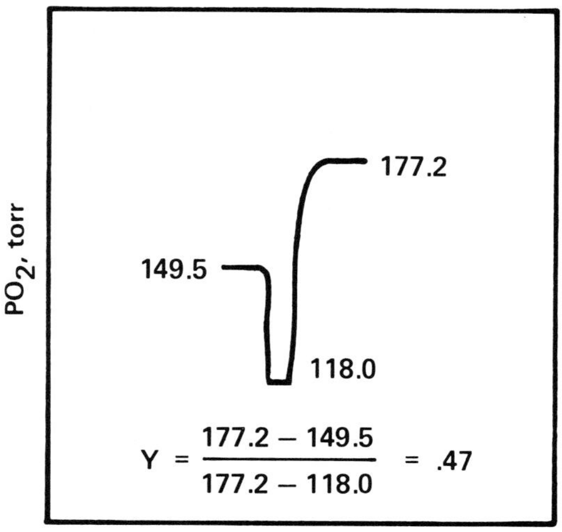

FIG. 5. Saturation measurement.

Mean corpuscular hemoglobin concentration increased in all subjects but remarkably so in one. Interestingly this individual also had a very high mean corpuscular hemoglobin concentration at sea level. Although he diligently consumed fluid, he may have become overly dehydrated. There was no correlation between mean corpuscular hemoglobin concentration and performance or acclimatization in the group as a whole.

Two previous expeditions to similar altitudes reported enough hematological data for detailed comparisons (Table 2; 4, 16). With a *t* test for small samples, we found that our red cell counts at 6,300 m are not different from those in 1973 but that our hemoglobin and hematocrit values are slightly different from the earlier data. Possible explanations for these discrepancies have been discussed (31); they are probably due to differences in experimental techniques and length of exposure to altitude.

TABLE 1. *Hematological data at various locations*

	Altitude				
	Sea level	5,400 m	6,300 m	8,050 m	8,848 m
No. of subjects	18	20	19	2	1
Hemoglobin, g/dl	14.5 ± 0.7	17.8 ± 1.0	18.8 ± 1.5	18.4	18.7
Hematocrit, %	43.8 ± 2.3	50.8 ± 2.6	53.4 ± 4.0	52.0	54.0
2,3-DPG/hemoglobin	0.838 ± 0.120	1.050 ± 0.090	1.040 ± 0.120	1.178	1.154
Base excess, meq/liter	−0.28 ± 1.22		−8.7 ± 1.7	−7.2	−5.9
pH, blood	7.399 ± 0.017		7.467 ± 0.034*	7.552	7.78†
P_{50}, mmHg	28.1 ± 1.1		27.6 ± 2.2	24.9	18.5
Pa_{O_2}, mmHg	90‡		39.1*		28†
P_{CO_2}, mmHg	40‡		18.4*	12.8†	7.5†
Sa_{O_2}, %	95‡		71		75

Values are means or means ± SE. 2,3-DPG, 2,3-diphosphoglycerate; Pa_{O_2}, partial pressure of O_2 in arterial blood; P_{CO_2}, partial pressure of CO_2; Sa_{O_2}, arterial O_2 saturation. * Arterial measurement. † See ref. 25. ‡ Assumed value.

TABLE 2. *Hematological data at extreme altitude*

	Altitude			
	5,970 m	5,350 m	5,400 m	6,300 m
Hemoglobin, g/dl	19.6 ± 1.0 (*n* = 24)	20.6 ± 0.4 (*n* = 10)	17.8 ± 1.0 (*n* = 20)	18.8 ± 1.5 (*n* = 19)
Hematocrit, %	55.8 ± 3.3 (*n* = 21)	63.8 ± 4.6 (*n* = 10)	50.8 ± 2.6 (*n* = 20)	53.4 ± 4.0 (*n* = 18)
Red blood cell count × 10^{-6}, fl	5.60 ± 0.43 (*n* = 18)	6.57 ± 0.68 (*n* = 10)		6.61 ± 0.40 (*n* = 18)
Corpuscular volume, fl	99.6	97.1		80.8
Corpuscular hemoglobin, pg	35.0	31.4		28.4
Corpuscular hemoglobin concentration, g/dl	35.1	32.3		35.2
Year (ref.)	1960/61 (16)	1973 (4)	1981	1981

Values are means or means ± SE. *n*, No. of subjects.

Acid-Base Status

Acid-base status was inferred by measuring pH of blood that had been equilibrated at known Pco_2 and Po_2. Saturation with oxygen had been measured as outlined in the previous section. The algorithms of Thomas (23) were used to calculate base excess, but no account was made for variation in temperature and the measured saturation was used rather than that estimated by Thomas (23). Direct estimation of base excess from arterial blood at Camp II demonstrated that the technique was valid (31).

Because direct measurements with arterial blood on the South Col or summit were not possible, in vivo pH was inferred from the tonometry data. Venous blood was sampled from two climbers on the South Col about 12 h after they returned from the summit. The blood was immediately placed on ice and taken to Camp II, where it was equilibrated to a final Pco_2 of 22.6 mmHg. The pH was measured, and base excess was calculated by using the Thomas algorithms (23). To estimate the summit pH in one of the climbers, we then used the alveolar Pco_2 (PA_{CO_2}) measured on the summit (7.5 mmHg; 25) and assumed $Pa_{CO_2} = PA_{CO_2}$. To construct an in vivo buffer line on the Siggaard-Andersen alignment nomogram (22), we reduced the hemoglobin concentration by one-third (21) and extrapolated to a summit pH of approximately 7.78.

Table 1 summarizes the findings at all altitudes. As altitude increases, metabolic compensation for respiratory alkalosis is progressively less effective. Our figures predict that for complete metabolic compensation on the summit, base excess would have to be about −18 meq/liter for blood pH to be 7.4. Thus blood pH was much higher than was predicted before the expedition (26).

Uncompensated respiratory alkalosis was found also in Andean natives of 4,250 m (13, 27) and in Sherpas at a similar altitude (13). Our studies and these previous results suggest that sojourners apparently do not compensate either, even after several weeks of acclimatization. This failure of blood pH to return to sea-level values is still not understood. Estimations of in vivo base excess and, even more, arterial plasma pH are uncertain for several reasons, and the in vivo pH-Pco_2 relationship is well known to differ from that found in vivo (21).

Blood Oxygen Affinity and 2,3-Diphosphoglycerate

The blood P_{50} values in Table 1 have been adjusted to in vivo pH assuming a Bohr factor of −0.45, corresponding to a 2,3-DPG:hemoglobin ratio of 1.25 (20). The P_{50} estimations from our previous work relating 2,3-DPG, pH, and Pco_2 (29) are 28.3 mmHg at sea level and 29.4 mmHg at Camp II, compared with mean measured values of 28.1 mmHg and 29.8 mmHg. Thus the P_{50} is completely accounted for by the known effectors of red cell oxygen affinity.

The 2,3-DPG:hemoglobin ratio increases in proportion to blood pH, in accordance with the view that intraerythrocyte pH is the chief controlling factor in 2,3-DPG synthesis (8). This compound strongly influences the position of the OEC by markedly reducing oxygen affinity.

However, increased pH by itself has the opposite effect; it increases oxygen affinity. The effect of 2,3-DPG in modulating oxygen affinity is limited, however, because it binds to deoxyhemoglobin in a molar ratio of 1, so that accumulating an excessive concentration has little effect (20). Perhaps for this reason the effect of alkalosis itself on the OEC under conditions of extreme hyperventilation is more important than that of 2,3-DPG.

In Vivo Blood Oxygenation

Figure 6 summarizes our findings on Mount Everest. As altitude increases, Pa_{O_2} decreases, resulting in a steady fall in arterial saturation. Because pH and 2,3-DPG increase together, there is little change in in vivo P_{50} up to an altitude of about 6,300 m. Above that altitude, however, uncompensated respiratory alkalosis dominates and presumably 2,3-DPG has little additional effect; the result is a marked fall in P_{50} and preservation of arterial saturation. Our calculations predict in fact that arterial saturation may actually be higher at 8,050 m and 8,848 m than at 6,300 m!

Climbers usually elect to acclimatize for 1 or 2 days below the summit, then attempt the summit in sudden pushes. The progressive shift to the left

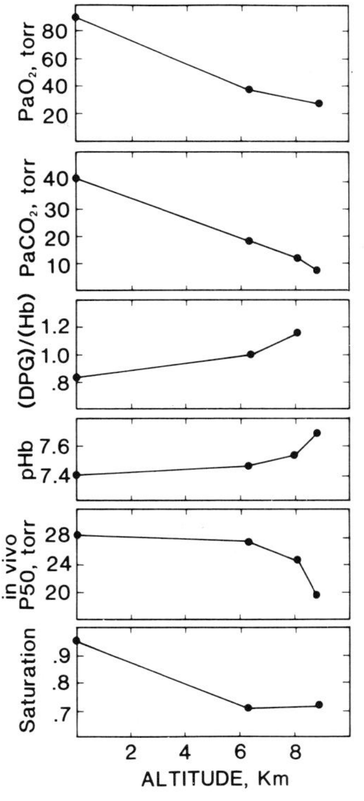

FIG. 6. Effectors of blood O_2 affinity. Net result (*bottom panel*) is a protected arterial O_2 saturation at extreme altitude.

of the OEC above Camp II suggests that this intuitive climbing strategy might be a good one; a left shift could confer some advantage because arterial saturation increases. Indeed experimental animals survive better at extreme altitude when the OEC is left shifted (6).

PHYSIOLOGICAL IMPORTANCE OF THE OXYGEN EQUILIBRIUM CURVE

Combining these results with our previous studies in the Andes (27), we formulated the hypothesis that in natives and sojourners the position and shape of the whole blood OEC is maintained within narrow limits by opposing cellular regulatory mechanisms up to altitudes between 4,000 m and 6,300 m. Above that range the effect of alkalosis prevails and the OEC shifts progressively to the left.

At first this increase in oxygen affinity at extreme altitude appears to benefit oxygen transport because, coupled with the extreme hyperventilation (25), it should preserve arterial oxygen saturation during severe hypoxia (31). However, an increased oxygen affinity also requires lower oxygen pressures in tissues to unload oxygen from hemoglobin. Unfortunately, much less is known about the process of unloading oxygen in tissues than about loading in the lung.

Direct experimental evidence for the positive or negative effects of shifts in the OEC in humans is lacking because no known method pharmacologically shifts the OEC without harmful or complicating physiological or biochemical effects. Perhaps the most relevant human observations were made by Hebbel and co-workers (9), who found that subjects with abnormally high oxygen affinity due to a genetic mutation of the hemoglobin molecule seemed to adapt better to moderate altitude than control subjects. However, such subjects differ in many ways from trained climbers. For example, their OEC is permanently left shifted (and biphasic!), so their adaptation to effective hypoxia is lifelong. Whether or not these subjects are similar to normal humans in ventilatory drives, tissue responses, red cell mass, or pulmonary circulation is not known.

Animal studies provide further evidence that a left shift of the OEC may be beneficial at extreme altitude. Eaton and co-workers (6) showed that rats with carbamylated hemoglobin had increased red cell oxygen affinity and improved survival at extreme altitude compared to appropriate control animals. Monge and Whittembury (15) pointed out that most animals native to high altitude (alpaca, vicuña, llama, chinchilla, vizcacha, yak, ostrich, and huallata) have increased blood oxygen affinity compared to their sea-level relatives (camel, rabbit, ox, and a variety of sea-level birds). This may indicate that humans have not had sufficient evolutionary time (generations) to adapt to high altitude, either as sojourners or natives. However, extrapolating from animals to humans in this instance is dangerous because many other physiological variables may be at least as important as the position of the OEC in adaptation to hypoxia.

Unfortunately the issue of the benefit of a right- or left-shifted OEC in human sojourners and high-altitude natives will not be completely settled until a safe way is found to shift the curve and observe its effects. Moreover any resultant shift in an individual must also be considered in the context of his unique combination of physiological factors, such as ventilatory mechanics, hypoxic drive, and hemoglobin concentration. In any case the results from the 1981 American Medical Research Expedition to Everest indicate that the body can maintain the OEC close to its sea-level position and shape up to 6,300 m altitude. Very possibly manipulating the OEC pharmacologically could improve performance at extreme altitude. A better understanding of the role of red cell oxygen affinity in the overall physiological response to hypoxia remains a worthwhile goal.

REFERENCES

1. BARCROFT, J., C. A. BINGER, A. V. BOCK, J. H. DOGGART, H. S. FORBES, G. HARROP, C. MEAKINS, AND A. C. REDFIELD. Observations upon the effect of high altitude on the physiological processes of the human body carried out in the Peruvian Andes chiefly at Cerro de Pasco. *Philos. Trans. R. Soc. London Ser. B* 211: 351, 1922.
2. BENESCH, R., AND R. E. BENESCH. The effect of organic phosphates from human erythrocyte on the allosteric properties of hemoglobin. *Biochem. Biophys. Res. Commun.* 26: 659–667, 1968.
3. BOLTON, W., AND M. F. PERUTZ. Three dimensional fourier synthesis of horse deoxyhemoglobin at 2A resolution. *Nature London* 228: 531, 1970.
4. CERRETELLI, P. Limiting factors to oxygen transport on Mount Everest. *J. Appl. Physiol.* 40: 658–667, 1976.
5. CHANUTIN, A., AND R. R. CURNISH. Effect of organic and inorganic phosphates on the oxygen equilibrium curve of human erythrocyte. *Arch. Biochem. Biophys.* 121: 96–102, 1968.
6. EATON, J. W., T. D. SKELTON, AND E. BERGER. Survival at extreme altitude: protective effect of increased hemoglobin-oxygen affinity. *Science* 85: 743–744, 1974.
7. FATT, I. *Polarographic Oxygen Sensor. Its Theory and Its Application in Biology, Medicine, and Technology.* Cleveland, OH: CRC, 1976, p. 49–57.
8. GARBY, L., AND J. MELDON. *The Respiratory Functions of Blood.* New York: Plenum, 1977, p. 196–199.
9. HEBBEL, R. P., J. W. EATON, R. S. KRONENBERG, E. D. ZANJANI, L. G. MOORE, AND E. M. BERGER. Human llamas. Adaptation to altitude in subjects with high hemoglobin oxygen affinity. *J. Clin. Invest.* 62: 593–600, 1978.
10. HILL, A. V. The possible effects of the aggregation of the molecules of hemoglobin on its oxygen dissociation curve. *J. Physiol. London* 40: 4–7, 1910.
11. HILL, R. Oxygen dissociation curve of muscle hemoglobin. *Proc. R. Soc. London Ser. B* 120: 472, 1936.
12. HURTADO, A. Animals in high altitudes: resident man. In: *Handbook of Physiology. Adaptation to the Environment,* edited by D. B. Dill and E. F. Adolf. Washington, DC: Am. Physiol. Soc., 1964, sect. 4, chapt. 54, p. 843–860.
13. LAHIRI, S., AND J. S. MILLEDGE. Acid-base in Sherpa altitude residents and lowlanders at 4880 m. *Respir. Physiol.* 2: 323–334, 1967.
14. LENFANT, C., AND K. SULLIVAN. Adaptation to high altitude. *N. Engl. J. Med.* 284: 1298–1309, 1971.
15. MONGE, C. C., AND J. WHITTEMBURY. Increased hemoglobin oxygen affinity at extremely high altitudes. *Science* 29: 843, 1974.
16. PUGH, L. G. C. E. Blood volume and haemoglobin concentration at altitudes above 18,000 ft (5500 m). *J. Physiol. London* 170: 344–354, 1964.
17. ROSSI-BERNARDI, L., M. LUZZANA, M. SAMAJA, M. DAVI, D. DARIVA-RICCI, J. MINOLI, B. SEATON, AND R. L. BERGER. Continuous determination of the oxygen dissociation curve for whole blood. *Clin. Chem. Winston-Salem, NC* 21: 1747–1753, 1975.
18. SAMAJA, M., A. MOSCA, M. LUZZANA, L. ROSSI-BERNARDI, AND R. M. WINSLOW. Equations and nomogram for the relationship of human blood P_{50} with 2,3-diphosphoglycerate, CO_2, and H^+. *Clin. Chem. Winston-Salem, NC* 7: 1856–1861, 1981.
19. SAMAJA, M., AND E. ROVIDA. A new method to measure the hemoglobin oxygen saturation by the oxygen electrode. *J. Biophys. Biochem. Methods* 7: 143–152, 1983.
20. SAMAJA, M., AND R. M. WINSLOW. The separate effects of H^+ and 2,3-DPG on the oxygen equilibrium curve of human blood. *Br. J. Haematol.* 41: 373–381, 1979.
21. SIGGAARD-ANDERSEN, O. An acid-base chart for arterial blood with normal and pathophysiological reference areas. *Scand. J. Clin. Lab. Invest.* 27: 240–245, 1971.
22. SIGGAARD-ANDERSEN, O., AND K. ENGEL. A new acid-base nomogram. An improved method for the calculation of the relevant blood acid-base data. *Scand. J. Clin. Lab. Invest.* 12: 177–186, 1960.
23. THOMAS, L. J. Algorithms for selected blood acid-base and blood gas calculations. *J. Appl. Physiol.* 33: 154–158, 1972.
24. WEST, J. B. Man at extreme altitude. *J. Appl. Physiol.: Respirat. Environ. Exercise Physiol.* 52: 1393–1399, 1982.
25. WEST, J. B., P. H. HACKETT, K. H. MARET, J. S. MILLEDGE, R. M. PETERS, JR., C. J. PIZZO, AND R. M. WINSLOW. Pulmonary gas exchange on the summit of Mount Everest. *J. Appl. Physiol.: Respirat. Environ. Exercise Physiol.* 55: 678–687, 1983.
26. WEST, J. B., AND P. D. WAGNER. Predicted gas exchange on the summit of Mt. Everest. *Respir. Physiol.* 42: 1–16, 1981.
27. WINSLOW, R. M., C. C. MONGE, N. J. STATHAM, C. G. GIBSON, S. CHARACHE, J. WHITTEMBURY, O. MORAN, AND R. L. BERGER. Variability of oxygen affinity of blood: human subjects native to high altitude. *J. Appl. Physiol.: Respirat. Environ. Exercise Physiol.* 51: 1411–1416, 1981.

28. WINSLOW, R. M., J. M. MORRISSEY, R. L. BERGER, P. D. SMITH, AND C. G. GIBSON. Variability of oxygen affinity on normal blood: an automated method of measurement. *J. Appl. Physiol.: Respirat. Environ. Exercise Physiol.* 45: 289–297, 1978.
29. WINSLOW, R. M., M. SAMAJA, N. J. WINSLOW, L. ROSSI-BERNARDI, AND R. I. SHRAGER. Simulation of the continuous blood O_2 equilibrium curve over the physiological pH, DPG, and P_{CO_2} range. *J. Appl. Physiol.: Respirat. Environ. Exercise Physiol.* 54: 524–529, 1983.
30. WINSLOW, R. M., M. L. SWENBERG, R. L. BERGER, R. I. SHRAGER, M. LUZZANA, M. SAMAJA, AND L. ROSSI-BERNARDI. Oxygen equilibrium curve of normal human blood and its evaluation by Adair's equation. *J. Biol. Chem.* 252: 2331–2337, 1977.
31. WINSLOW, R. M., SAMAJA, M., AND WEST, J. B. Red cell function at extreme altitude on Mount Everest. *J. Appl. Physiol.: Respirat. Environ. Exercise Physiol.*, 56: 109–116, 1984.

7

Sleep and Periodic Breathing at High Altitude: Sherpa Natives Versus Sojourners

SUKHAMAY LAHIRI, KARL H. MARET, MINGMA G. SHERPA, AND RICHARD M. PETERS, JR.

Department of Physiology, Institute for Environmental Medicine, University of Pennsylvania School of Medicine, Philadelphia, Pennsylvania

FACTORS CONTRIBUTING to periodic breathing at high altitude during sleep or somnolence are the subject of this chapter. The Sherpa high-altitude natives in the Himalayas who were volunteers for a recent study receive special attention because the particular characteristics of their respiratory control systems may offer crucial tests for a general model of periodic breathing.

Adult natives of high altitude in various parts of the world show a more fully compensated respiratory alkalosis (6, 25, 30, 38, 46). Also they manifest an attenuated ventilatory response to hypoxia in the waking state (see chapt. 13; 36, 38, 47, 51). This is intriguing because of their more acid internal milieu of peripheral and central chemoreceptors. Nonetheless a relatively insensitive chemoreflex would dampen oscillations of the control system. On the other hand, a greater acidity may not allow apnea to develop. As a result these subjects may exhibit a reduced trend of periodic breathing with apnea at high altitude. Confirmation of these predictions would support the initial criteria that the subjects have a blunted hypoxic ventilatory drive and a more compensated respiratory alkalosis. These, however, would not exclude other contributing factors.

There are several forms of periodic breathing; some investigators have used the term *Cheyne-Stokes breathing* to describe any one of them. Confusion is therefore expected. In this article we use the term *periodic breathing* to describe a repetitive long cyclic pattern of breathing with or without apnea. With apnea we qualify it accordingly. Thus apnea may appear or disappear, but cycles may persist.

BACKGROUND

Periodic Breathing

One of the earliest observations on periodic breathing at high altitude was made in 1886 by Mosso (39; see also 26, 54). The records of thoracic respiration that he made on his brother at 3,620 m (Ginifetti Hut) and 4,650 m (Regina Margherita Hut) are reproduced in Figure 1. The periodicity particularly of tidal volume is obvious at 3,620 m, but it developed a "morbid intensity" at 4,560 m, whereupon periodic apnea of 12 s appeared between clusters of a few forceful breaths. These observations also indicate that the intensity of hypoxia increased the magnitude of periodic breathing. He compared these breathing patterns with those found in patients with circulatory problems described by Cheyne (12) and Stokes (50). The factors contributing to this type of periodic breathing have been investigated both experimentally and theoretically. Douglas and Haldane (16) demonstrated that periodic breathing could be produced experimentally in normal waking persons by want of oxygen caused by subnormal oxygen pressure, by circulatory deficiency, or by a combination of both. Douglas et al. (17) later investigated periodic breathing on Pikes Peak (4,300 m) and confirmed the prediction that oxygen want and hypocapnia were critical factors. They also noted sleep disturbances, an observation that has been amply confirmed but not adequately characterized since.

At the time of this work, arterial chemoreceptors had not been discovered. The effect of oxygen want was attributed to cerebral lacticacidosis; this has

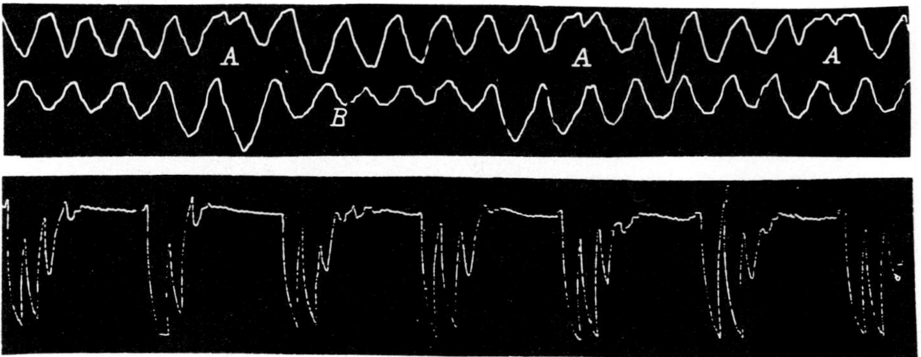

FIG. 1. Earliest published record of periodic breathing pattern of a subject at 3,620 m (*upper panel*) and 4,560 m (*lower panel*) during somnolence or sleep. [From Mosso (39).]

been proved incorrect. The discovery of oxygen sensors in the carotid and aortic bodies (13, 24) introduced a new element that was not appreciated until many years later. Pryor (43), from his observations of human patients, and Guyton et al. (23), from dogs, marshaled evidence favoring a circulatory delay between the lungs and respiratory center that shifted the temporal feedback relationship between the plant (lungs) and the controller (respiratory center). They suggested that this resulted from a delayed arrival of the chemical signal from the lungs to the controller in the central nervous system and thereby caused periodic breathing. The required delay in the experiments of Guyton et al. (23) was too long, however, and both studies appeared to omit the role of peripheral chemoreceptors. Lange and Hecht (35) further examined Cheyne-Stokes breathing in patients with heart disease and discounted hypoxemia but suggested that a phase lag between lungs and brain was the primary mechanism. In this model the controller responded to partial pressure of carbon dioxide (P_{CO_2}) alone. The study of Lange and Hecht was done in Salt Lake City, Utah, at an altitude of 1,250 m, which is mildly hypoxic. The arterial P_{CO_2} (Pa_{CO_2}) data provided by Lange and Hecht (35) do show hyperventilation in both normal subjects and patients (32 vs. 27 mmHg).

In contrast to the foregoing observations, Anderson et al. (1) emphasized arterial chemosensory input as the primary cause of circulatory (Mayer waves) and respiratory instability after hemorrhagic hypotension in anesthetized dogs. Preiss et al. (42) reported that periodic breathing was associated with Mayer waves but that it persisted in some instances even when the arterial blood pressure oscillations were eliminated. They concluded that the periodic-breathing pattern is not the result of an analogous pattern in the discharge of gas tension–sensitive or blood flow– and pressure-sensitive receptors fed back to the central nervous system but is generated within the central nervous system. Brown and Plum (7) also showed that Cheyne-Stokes breathing in humans could originate from within the central nervous system. These results showing a neurogenic basis could be attributed to an abnormally high gain of the system controlling breathing.

Cherniack and colleagues (10, 11) studied breathing instability both experimentally and theoretically and emphasized the characteristics of the breathing system (e.g., alinearities in controller gain, thresholds, circulation time, and gas stores) that contribute to the instability. More recently Khoo et al. (27) emphasized the role of arterial chemoreceptors in short-term instability of breathing. Brusil et al. (8), describing a filtering method to identify respiratory oscillations, found that breathing oscillations existed at a higher altitude in the awake state. They also noted that the oscillations were most prominent after acclimatization at 3,050 m and were not a function of acute change in acid-base balance.

General Model

The foregoing brief review of the literature makes it clear that periodic breathing may develop for many reasons, all related to various components of the chemoreflex pathways. Therefore one explanation may not fit all observations. However, the primary factors that could contribute to periodic breathing at high altitude are *1*) hypoxemia, which increases gain of the stimulus-response relationship of peripheral chemoreceptor activity; *2*) respiratory alkalosis, which diminishes central chemosensory drive and makes respiration more dependent on peripheral chemosensory input but also decreases the gain of hypoxic drive; *3*) increased oscillations of alveolar gases, particularly carbon dioxide, determined by expanded lung air due to hypobaria and increased alveolar ventilation; and *4*) increased central carbon dioxide sensitivity (37). Sleep or somnolence would provide the following additional factors: *1*) an increased central respiratory threshold, particularly for carbon dioxide and hydrogen ion drives (9); *2*) an associated acutely reduced carbon dioxide sensitivity (45); *3*) a decreased "waking" drive; and *4*) a possible enhanced circulatory delay between lungs and controller, which consists of chemosensors in the carotid and aortic bodies and respiratory neurons in the brain stem. Obviously the combination and interaction of these factors are also important determinants.

Acclimatization to a given hypoxic environment would reduce alkalosis and improve Pa_{O_2}. The latter and a reduction of central alkalosis would reduce the chances of periodic breathing with apnea at a moderate altitude but may not do so at a higher altitude, which provides a stronger hypoxic stimulus.

Whatever general model develops for the sojourners at high altitude should apply to the natives of high altitude as well. Any exception to the rule must then be attributed to the special known characteristics of the control system or to unknown physiological properties in the natives of high altitude.

EXPERIMENTS

Respiratory studies during sleep were carried out on adult male volunteers at the Everest Base Camp Laboratory (5,400 m) of the 1981 American Medical Research Expedition to Everest. The laboratory was housed in a prefabricated hut, which was heated to a temperature between 15°C and 20°C. The geographical setting is shown elsewhere in this volume. The barometric pressure (mean ± SE) was 400.4 ± 2.7 mmHg. The volunteers were adult males (age 20–48 yr), including six Sherpas born and continually residing above 3,500 m, one low-altitude Sherpa born and residing at or below 1,220 m, and seven low-landers from near sea level.

For the sleep studies, ventilation was measured by inductive plethysmography, arterial oxygen saturation (Sa_{O_2} %) by ear oximetry, heart rate by electrocardiography (ECG), and sleep-state index by electrooculography (EOG; although not by itself a specific probe). Alveolar gases were changed by administering nitrogen, oxygen, or carbon dioxide through the nasal prongs to the inspired air. Inspired PO_2 and PCO_2 (PI_{O_2} and PI_{CO_2}) were noted. Occasionally the end-tidal values were also measured.

Waking ventilation sensitivity to hypoxia was measured by nitrogen or oxygen tests of Dejours (14), as adapted by Edelman et al. (19). Additional measurements were made by carbon dioxide–rebreathing tests at two levels of Pa_{O_2} (>200 mmHg and ~50 mmHg), a modified version of Read's method (44).

SLEEP APNEA IN LOWLANDERS AT 5,400 METERS

Periodic Breathing and Oxygen

The breathing patterns during sleep of a lowlander who resided several weeks at 5,400 m are shown in Figure 2. When this experiment began the subject had been at 5,400 m for 5 wk. While breathing air (PI_{O_2} 73 mmHg) the subject showed periodic breathing with apnea throughout sleep. A cluster of three or four forceful breaths in ~10 s was followed by apnea of about the same duration. The apnea was "central" because both abdominal and rib cage movements were absent. Apnea was followed first by inspiration and then expiration. After the breaths Sa_{O_2} % increased, and in the middle of apnea it began to decrease. This out-of-phase relationship between breathing pattern and Sa_{O_2} % is expected because peak ventilation raised Pa_{O_2} and lowered Pa_{CO_2}, both of which increased Sa_{O_2} %. The time lag between peak ventilation and peak saturation (indicating circulation time) was ~12 s. This lung-ear circulation time in men during sleep in the supine position at 5,400 m is longer than the 6.8–9.4 s found during rest at sea level (50). The total cycle length was ~20 s, nearly half of which was the delay time between lung and ear, that is, between left heart and carotid bodies. A similar pattern of periodic breathing in lowlanders was recorded at the Camp II Laboratory (6,330 m) during the expedition (J. B. West, personal communication).

Adding oxygen to the inspired air raised Sa_{O_2} % and then immediately increased the apneic period to 17 s with only slight increases in cycle duration and lung-ear circulation time. These results indicated a loss of hypoxic ventilatory drive with the loss of hypoxemia. Because of low oxygen stores,

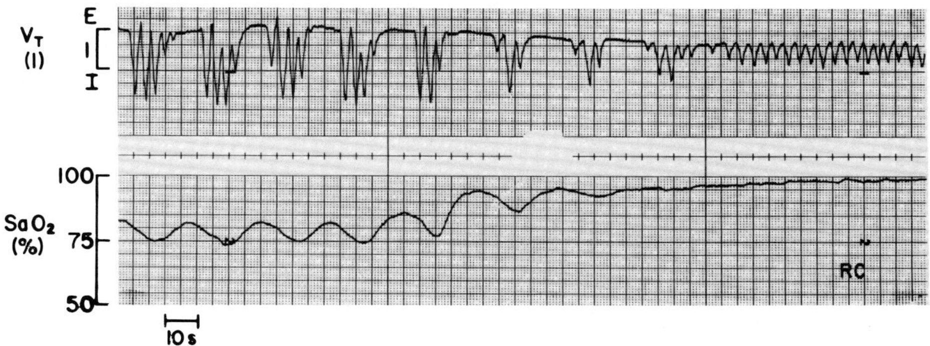

FIG. 2. Periodic breathing in lowlander at 5,400 m during sleep and the effect of transition from air to O_2 breathing. *Tracings* from *top*: tidal volume (VT, inspiration downward) and arterial O_2 saturation (Sa_{O_2} %). Sudden rise in Sa_{O_2} % indicates O_2 administration.

Pa_{O_2} increases rapidly after oxygen administration, whereas Pa_{CO_2} rises slowly because of the characteristics of carbon dioxide stores in blood and other tissues. A small decrease in cardiac output, as indicated by a decreased heart rate, may have lengthened the circulation time. The profound effect of oxygen is consistent with the high hypoxic drive of this subject (see VENTILATORY SENSITIVITY TO HYPOXIA, p. 84). Four cycles after oxygen breathing and an elapse of 105 s, the apneic period shortened and a shallow rhythmic breathing resumed. The circulatory delay alone during hyperoxia therefore did not cause a profound periodic breathing.

The immediate effect of oxygen breathing could be attributed to the peripheral chemoreflex. This effect is best explained with an illustration of carotid body chemosensory activity and the breathing pattern during transition from hypoxia to hyperoxia in a cat (Fig. 3). After a breath of oxygen, carotid chemoreceptor activity abruptly disappeared, inspiratory volume decreased, and the expiratory duration increased, resembling apnea. During the latter response, carotid chemoreceptor activity began to increase and a breath appeared that again sharply decreased the chemoreceptor activity. The cycles continued, and the breath intervals gradually decreased. When the chemosensory inputs are interrupted by cutting the carotid sinus and aortic nerves, the immediate ventilatory responses of oxygen breathing disappear. This indicates that the immediate direct effect of oxygen on cerebral blood flow and the consequent indirect effect on ventilation were not significant.

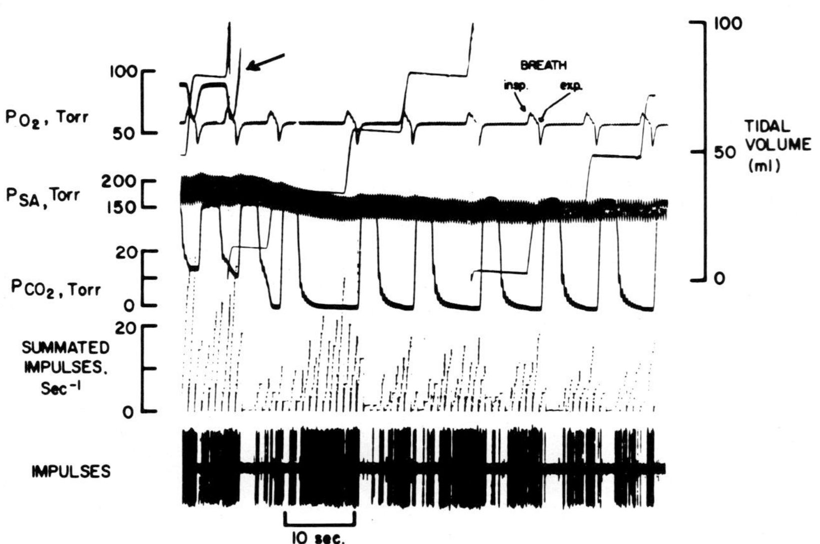

FIG. 3. Effect of transition from air to O_2 breathing on carotid chemoreceptor activity and ventilation in an anesthetized cat at sea level. End-tidal partial pressure of CO_2 (P_{CO_2}) was held constant. Oscillations of chemoreceptor activity with breaths are clear. *Tracings* from *top*: tracheal P_{O_2} (*arrow*), inspiratory tidal volume (*steps*), tracheal air flow, systemic arterial blood pressure (P_{SA}), tracheal P_{CO_2}, chemoreceptor discharge rate, and impulses.

Initiation of Periodic Breathing

In most of our lowland subjects periodicity in breathing did not disappear totally during oxygen-enriched air breathing (see, e.g., Fig. 4). A periodicity of ~25 s can be seen at an Sa_{O_2} of 98%. Resumption of air breathing was followed by a decrease in Sa_{O_2} % and an increase in ventilation. During this transition the periodicity seemed to disappear until the mean Sa_{O_2} reached 82%, whereupon the periodicity developed an apneic pattern.

The cycle duration with air breathing was ~21 s, half of which was an apneic period. The time lag between peak ventilation and peak Sa_{O_2} % was ~12 s. These characteristics of periodic breathing are similar to those in another subject shown in Figure 2.

The EOG indicated quiet sleep during oxygen-enriched air breathing. In the transition to air breathing there was a short burst of movement, indicating rapid-eye-movement (REM) sleep, which may have contributed to the brief period of an irregular breathing pattern (22, 39, 40).

There are claims in the literature that periodic breathing disappears with oxygen breathing (11, 17). An example from Douglas et al. (Fig. 5; 17) shows that periodic breathing was clearly attenuated by oxygen breathing, but it certainly did not disappear. The cycle time was lengthened, however. Thus the claim is not substantiated. These long cycles, however, are not visible in the awake state, although other analytical methods might detect them (8).

One of the lowlanders did not develop periodic breathing with apnea at any time throughout the night, although he did show small oscillations and irregularities in breathing. Oxygen breathing caused little effect. He also had a little ventilatory sensitivity to hypoxia in the waking state (see VENTILATORY SENSITIVITY TO HYPOXIA, p. 84).

Effect of Carbon Dioxide Inhalation

Because of hypocapnia and alkalosis at altitude, the role of carbon dioxide was investigated by allowing the subject to inhale carbon dioxide in air (Fig. 6). Before carbon dioxide inhalation the cycle time consisted of 20 s with apnea of 10 s. The interval between peak ventilation and peak Sa_{O_2} % was ~11 s. Inhalation of carbon dioxide promptly eliminated apnea, but periodic breathing with the following characteristics continued: *1*) fluctuations of the expiratory period—longer period with smaller volume and shorter period with larger volume; *2*) reduction of the cycle time (to ~17 s); and *3*) constant interval of ~11 s between peak ventilation and peak Sa_{O_2} %. Thus carbon dioxide inhalation did not fundamentally change the periodic property. This periodicity was also evident in the corresponding oscillations of Sa_{O_2} %, but its amplitude diminished partly because of lack of apnea and increased Pa_{O_2}, as seen in higher mean Sa_{O_2} %, and also because of dampened Pa_{CO_2} oscillations resulting from carbon dioxide inhalation. A higher level of inhaled carbon dioxide would further dampen the oscillation. Also a longer duration of carbon dioxide

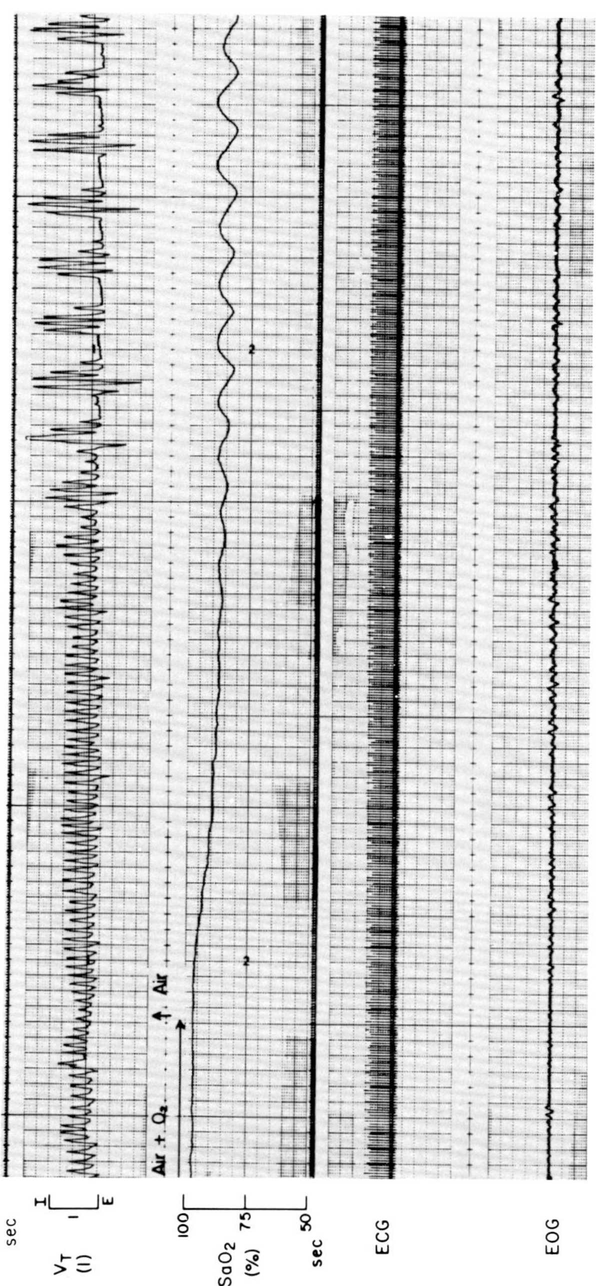

FIG. 4. Changes in breathing pattern during transition from O_2 to air breathing in a lowlander at 5,400 m during sleep. Inspiration is upward for V_T. ECG, electrocardiograph; EOG, electrooculogram.

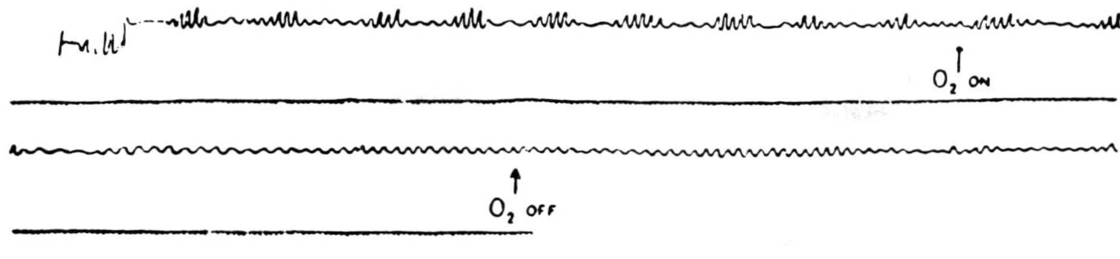

FIG. 5. Breathing oscillations in a lowlander during sleep at 4,300 m showing the effect of O_2 breathing. *Tracings* from *top*: respiratory trace, time marker in s, respiratory trace, time marker, respiratory trace. [From Douglas et al. (17).]

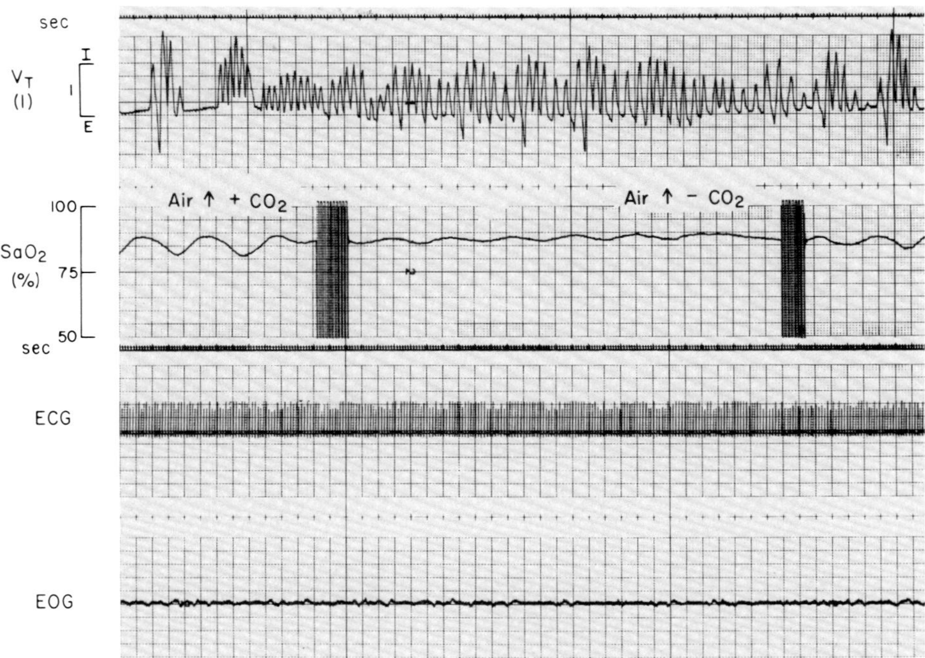

FIG. 6. Effect of inhalation of CO_2 on the periodic breathing of a lowlander during sleep at 5,400 m.

inhalation to a steady state would stimulate breathing further, but that would not alter the periodic pattern of breathing.

After carbon dioxide was withdrawn from the inspired air, apnea promptly reappeared, which supports the conclusion that Pa_{CO_2} oscillations and the associated chemosensory oscillations play a critical role. The central oscillation of carbon dioxide and hydrogen ions probably made only a small contribution to the respiratory oscillations because of a long time constant of the central chemosensory system (46), but central alkalosis certainly contributed to apnea.

LACK OF SLEEP APNEA

High-Altitude Sherpa

Unlike lowlanders, none of the Sherpa high-altitude natives who camped and acclimatized at 5,400 m developed a sustained periodic breathing with apnea during sleep. Figure 7 shows their general pattern of breathing. This

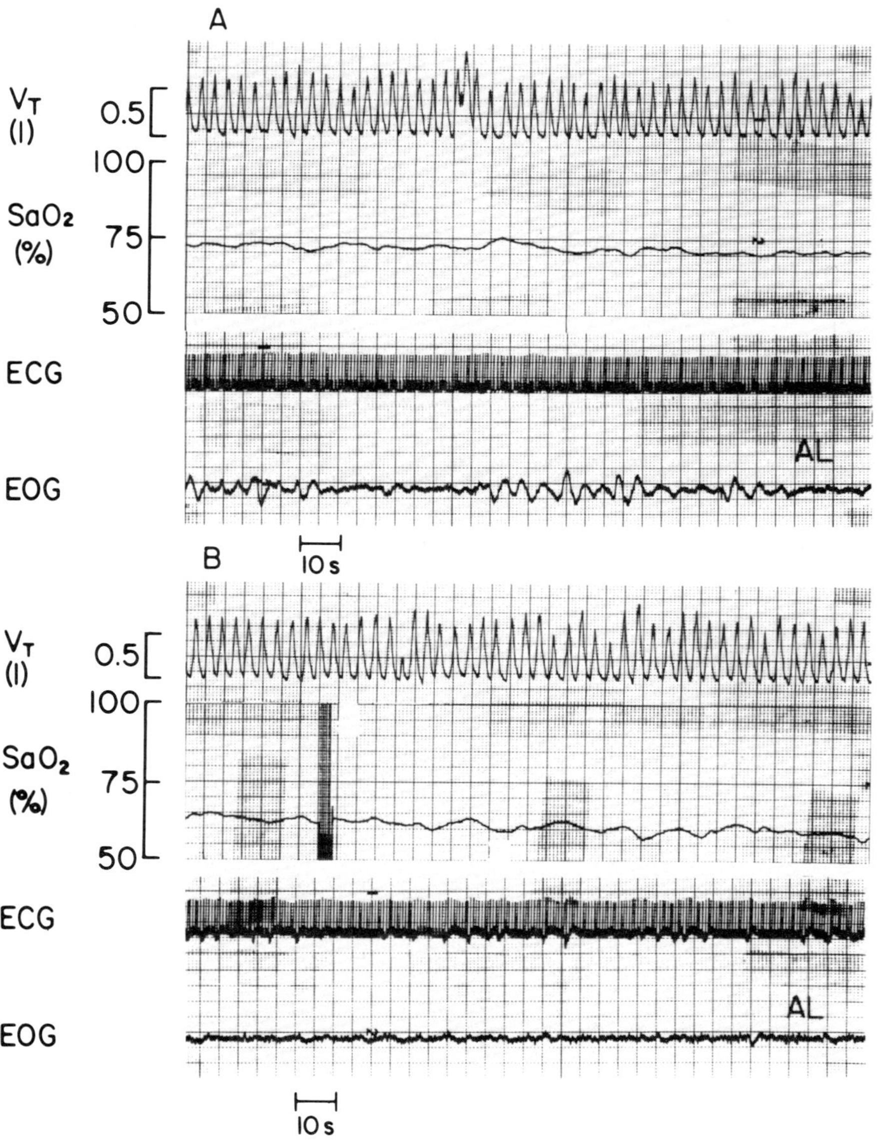

FIG. 7. Respiration in a Sherpa highlander during sleep at 5,400 m. Adding N_2 to inspired air lowered Sa_{O_2} % in *B*.

subject had an Sa_{O_2} % that varied between 71% and 75%. The EOG indicates REM sleep. Because hypoxia was the critical stimulus for periodic breathing with apnea in the sojourners during sleep at 5,400 m, it was thought that a greater hypoxic stimulus may cause a typical periodic breathing. Decreasing Sa_{O_2} % to 56–60% by lowering PI_{O_2} failed to produce such a result. The subject was not awake.

At another time, raising PI_{O_2} and Sa_{O_2} % by oxygen inhalation was followed by a transient decrease in tidal volume (Fig. 8). As Sa_{O_2} % approached 100%, ventilation increased. Both responses indicate little ventilatory sensitivity to hypoxia, as was found also in the waking state (see VENTILATORY SENSITIVITY TO HYPOXIA, p. 84).

Occasionally after a large breath at a low PI_{O_2} Sherpa highlanders also showed apnea and a few cycles of oscillations in breathing and in Sa_{O_2} %. An example is shown in Figure 9. The cycle time was ~14 s, 6 s shorter than the sustained cycles observed in the lowlanders, and the initial apneic period was ~8 s.

Low-Altitude Sherpa

A low-altitude Sherpa subject showed spontaneous periodic breathing with apnea, which appeared similar in most measured characteristics to those in the lowlanders (Fig. 10). The cycle time was 18–19 s with apnea for 9–10 s. The lung-to-ear delay was 12–13 s. While breathing oxygen-enriched air his

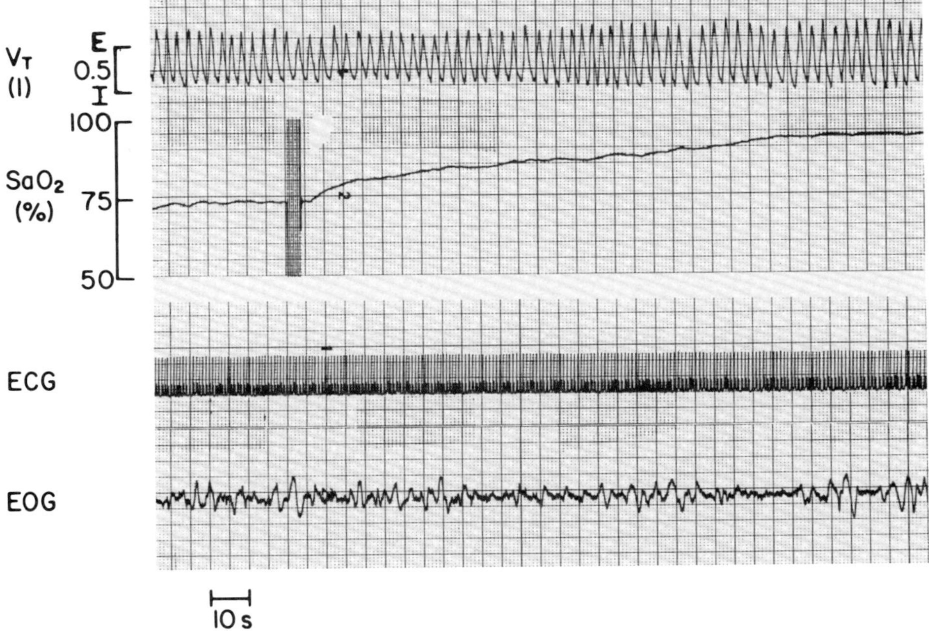

FIG. 8. Effect of raising the pressure of inspired O_2 on ventilation during sleep at 5,400 m in a Sherpa.

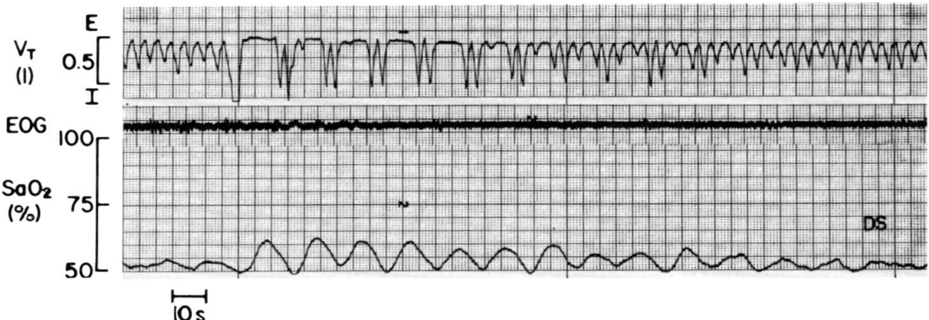

FIG. 9. Spontaneous decrease of Sa_{O_2} % during sleep in a Sherpa subject at 5,400 m. Breathing oscillation started after a deep breath, but it was not sustained. The EOG record is taken from strip chart of a 2nd recorder, which was run concurrently. [From Lahiri et al. (31).]

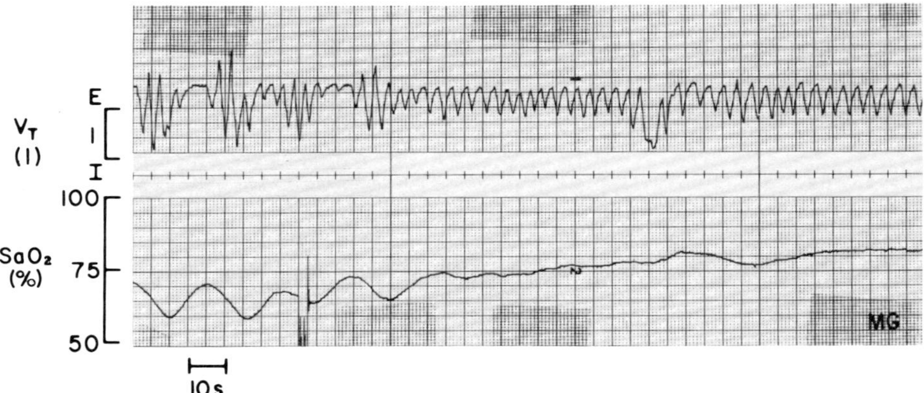

FIG. 10. Effect of a spontaneous large breath on breathing pattern during sleep at 5,400 m with O_2 breathing. Administration of O_2 started at break of Sa_{O_2} % trace.

large breathing cycles diminished in intensity and he resumed normal breathing unusually promptly. A large breath during oxygen inhalation, as with air breathing, was not followed by apnea or successive oscillations. These observations indicate that in the absence of hypoxic drive a perturbation of alveolar P_{CO_2} ($P_{A_{CO_2}}$) and Pa_{CO_2} was not enough to cause sufficient breathing oscillation. That is, the peripheral chemosensory input seems to be critical for the large oscillations. The results also suggest that Sherpa ancestry alone is not sufficient to prevent instability of breathing at high altitude.

VENTILATORY SENSITIVITY TO HYPOXIA

Ventilatory sensitivity to hypoxia was measured in all subjects of sleep studies during their waking states. The ventilatory responses corresponding to the minimal Sa_{O_2} % after nitrogen breaths were measured. These pairs of data were used to construct ventilatory response curves for each subject, the slopes of which provided indices for the ventilatory response to hypoxia. These representative data are given in chapter 13. The Sherpas showed a small slope as well as a lower ventilation in the hypoxic range. During transient hyperoxia, ventilation changed little in the Sherpa subject, whereas it decreased significantly in the lowlander.

SLEEP APNEA AND VENTILATORY SENSITIVITY TO HYPOXIA

The relationship between ventilatory sensitivity to hypoxia and the frequency of apnea is shown in Figure 11. The Sherpa highlanders who showed attenuated ventilatory responses to hypoxia in the awake state exhibited the least apnea during sleep. A lowlander in this group showed blunted ventilatory sensitivity to hypoxia. On the other hand, those with greater ventilatory sensitivity to hypoxia also manifested greater apneic frequency. Six of these seven subjects were lowlanders, and one was a Sherpa by birth but a lowlander by residence. Thus periodic breathing with apnea was strongly correlated with ventilatory sensitivity to hypoxia ($r = 0.85$) regardless of ancestry.

CARDIORESPIRATORY OSCILLATIONS

The periodic breathing with apnea accompanied oscillations of heart rate. It decreased after apnea and accelerated with the initiation of breathing. This fluctuation in the heart rate would probably result in a fluctuation of pulmonary blood flow, which in turn could influence peripheral chemoreceptor activity. A sufficiently low flow, although unlikely, might stimulate the peripheral chemoreceptors (34). On the other hand, a decreased blood flow could decrease PA_{CO_2} and raise PA_{O_2}; a corresponding change in the arterial blood gases would decrease the carotid body chemoreceptor activity and hence ventilation. An increase in blood flow would reverse the stimulus and response. Thus the circulatory fluctuation would reinforce the ventilatory periodicity.

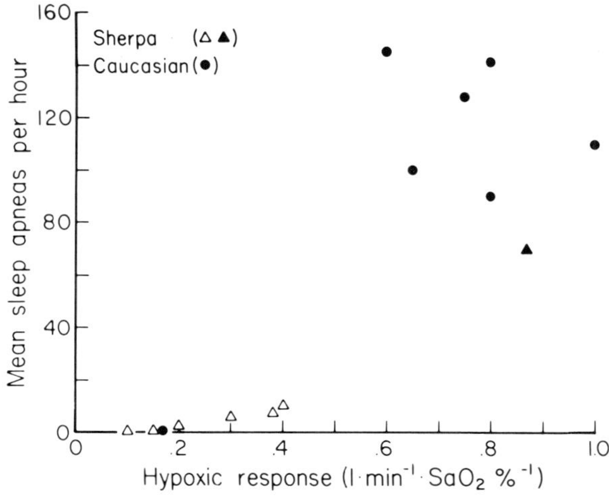

FIG. 11. Relationship between ventilatory hypoxic response and frequency of sleep apnea in Caucasian lowlanders and Sherpas at 5,400 m. [From Lahiri et al. (31).]

CYCLE AND LUNG-EAR CIRCULATION TIMES

The seven subjects who showed periodic breathing with apnea yielded the following data (means ± SD): total cycle time, 20.5 ± 1.8 s; apneic period, 10.2 ± 0.9 s; and lung-ear circulation time, 13.7 ± 1.4 s. The lung-ear circulation time is approximate because it neglects a possible delay in the oximeter reading. The cycle time is significantly greater than that predicted by Khoo et al. (27) for 5,400 m. Also the circulation time between the lungs and chemosensors was almost double the normally expected value (51). These timings appear favorable for a perpetual periodicity.

MECHANISMS

The observations strongly implicate peripheral chemoreflex with respiratory alkalosis as the basis of periodic breathing with apnea in the high-altitude sojourners; they also suggest that an attenuated chemoreflex and a more fully compensated respiratory alkalosis are responsible for a general lack of the phenomenon. A finite chemosensory input is needed to exceed the central respiratory threshold (20, 32). The observations indicate that this threshold is exceeded by a tonic input in the Sherpas, but an oscillatory signal or its effect is attenuated along with the mean hypoxic drive.

The rate sensitivity of peripheral chemoreceptors is relevant here. The carotid body chemoreceptors respond to a Pa_{CO_2} with an overshoot and to a decrease with an undershoot (2, 4). Hypoxia augments these dynamic responses to carbon dioxide (33), as shown in Figure 12. Ventilation follows such chemosensory input (5).

According to these properties of peripheral chemoreceptors, a small perturbation in breathing can initiate oscillations in the carotid chemosensory input, which in turn could reinforce it and perpetuate the cycle. This effect of the arterial chemical stimuli could be aided by a decreased lung volume and hence carbon dioxide store in the recumbent position during sleep. Also a reduced number of gas molecules in a given lung volume at high altitude would favor greater oscillations of Po_2 and Pco_2 with ventilation. The apnea after large breaths indicates not only a strikingly diminished peripheral chemosensory input but also a central chemosensory input inadequate for sustaining ventilation (20, 32). Presumably the residual uncompensated respiratory alkalosis in the lowlanders assists this (29). Pulmonary vagal reflexes may play a minor role (41).

Because of these physiological factors, the requirements for large oscillations in breathing seem to be set at high altitude. Sleep or somnolence provides the critical elements that force the initiation of periodic breathing with apnea. An increase of the central Pco_2 threshold makes the system more dependent on the peripheral drive. The controller then oscillates with the oscillations of the peripheral chemosensory input. Cherniack et al. (11) have reproduced this in experimental animals with cold block of the central chemosensory mecha-

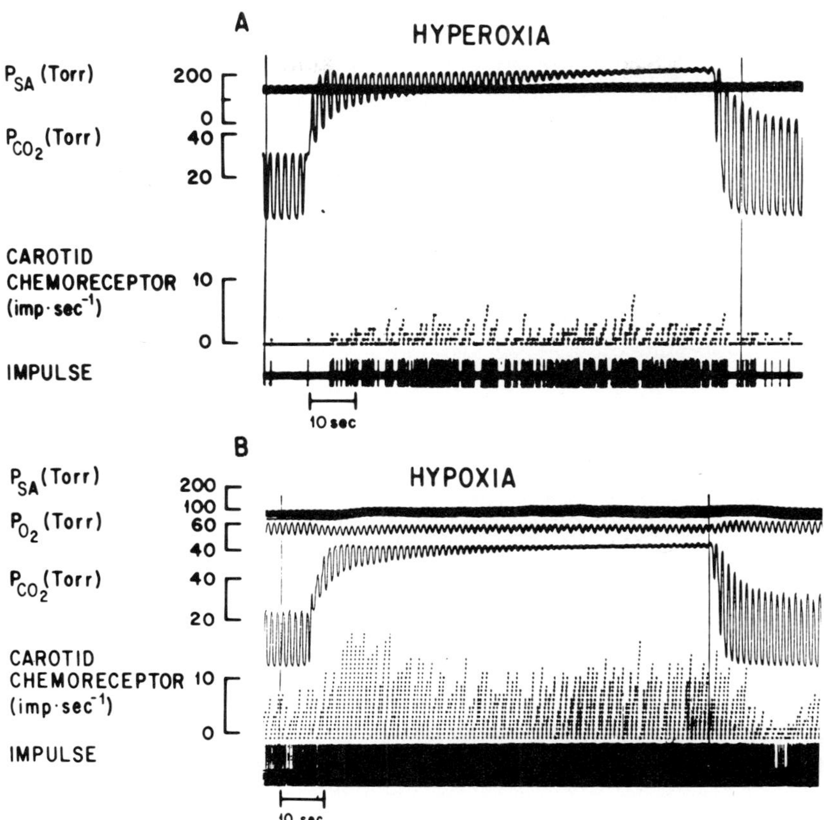

FIG. 12. Overshoot and undershoot in responses of carotid body chemoreceptors to Pa_{CO_2} changes in hyperoxia (*A*) and hypoxia (*B*) in anesthetized cat. *Tracings* from *top* in *A*: systemic arterial pressure (PSA), airway P_{CO_2}, carotid body chemoreceptor activity (summated impulse/s and impulse); in *B* airway P_{O_2} trace is shown also. Hypoxia augmented ON and OFF as well as steady-state responses. [From Lahiri et al. (33).]

nism. However, for this mechanism to work during sleep the block has to develop between the central chemosensors and the central respiratory neurons without compromising the effect of the peripheral chemosensory input. Decreasing ventilatory sensitivity to carbon dioxide and increasing P_{CO_2} threshold (9, 45) during sleep would similarly make the control system more dependent on the peripheral chemosensory input. A decreased hypoxic chemoreflex at the same time would, however, reduce the probability of periodic breathing with apnea. Until recently it was thought that hypoxic chemoreflex is not diminished during sleep (41). Douglas et al. (18) recently reported a reduction of hypoxic chemosensitivity particularly during REM sleep. However, according to our observations (Fig. 11), the magnitude of this reduction may not be sufficient to prevent periodic breathing with apnea.

From the mechanical point of view of bioengineering, the criteria for the control instability are met at high altitude and it is no wonder that breathing oscillations are prevalent in the sojourners. What prevents it in the high-altitude natives requires additional explanation.

Whether the diminished chemoreflex response is caused by a diminished chemosensory input is not clear. The lack of an appropriate chemosensory input may originate at the blood-gas stimulus level. If the functional residual capacity in the supine position is significantly greater in Sherpas than lowlanders (15, 29), the oscillations of respiratory gases would be more damped and reduce the chances of stimulus oscillations. At an increased hypoxic stimulus level a small perturbation of blood gases would cause a relatively large chemosensory response; such a response occurred in the Sherpas, but it was not sustained. Thus it appears that the respiratory oscillations were centrally suppressed.

One or more neurotransmitters in chronic hypoxia may play a role in this suppression. For example, catecholamine metabolism in the carotid body is strikingly increased during chronic hypoxia (34). These catecholamines can alter chemoreflex responses, but the changes occur within a few weeks of chronic hypoxia. Therefore they cannot account for the difference between sojourners and natives at high altitude. A decrease in the cellular receptor sites, however, could occur over a longer period because of a prolonged increased level of a neurotransmitter such as dopamine. However, our recent observations that carotid body chemosensory responses in the chronically hypoxic cat are not attenuated (3; also S. Lahiri, M. Pokorski, N. J. Smatresk, P. Barnard, and A. Mokashi, unpublished observations) militate against this reasoning. We have also found recently (49) that central blocking of dopamine receptor sites by haloperidol dramatically attenuates the steady-state ventilatory peripheral chemoreflex response without significantly affecting the central carbon dioxide and hydrogen ion responses, although the peripheral chemosensory responses are augmented. This observation indicates a potential mechanism for attenuation of hypoxic chemoreflex response.

A different sleep pattern probably was not responsible for the differences between the Sherpas and sojourners, because periodic breathing occurs during both REM and non-REM sleep (41) and even without actual sleep; also the sleep patterns in the high-altitude residents are not strikingly different (53). It must be emphasized that hypoxia profoundly alters sleep, which in turn affects respiration (39, 40). These aspects in the Andean and Himalayan high-altitude natives have not been fully investigated.

According to the hypothesis favored in this paper, periodic breathing may not develop until several days after birth, when peripheral chemoreflexes develop more fully (28, 29). Although this may be true in some infants, the mechanism of periodic breathing in newborns seems to lie in the central nervous system (21).

Young children native to high altitude, unlike adults, may show periodic breathing during sleep because their chemoreflex is like that of the normal sojourners. On the other hand, the patients who show blunted ventilatory sensitivity to hypoxia may not manifest periodic breathing with apnea. These phenomena require further studies.

SIGNIFICANCE

If homeostasis and stability are goals of the integrative physiological system, then periodic breathing interrupted with apnea is not desirable. However, the control system has the built-in potential for large oscillations that develop in the high-altitude sojourner. In the realm of physiological adaptation and evolution, we have witnessed time and again that a response is modified as the need arises. In high-altitude dwellers this modification has taken place apparently at the expense of a brisk ventilatory response to hypoxia. Living systems consist of a series of trade-offs; in this instance the trade-off lies between a sensitive chemoreflex response to hypoxia and stable breathing at high altitude.

We are indebted to all members of the 1981 American Medical Research Expedition to Everest, especially to J. B. West, the leader, and to volunteers for the study. We owe gratitude to S. Girme, our Sherpa leader, for his help with the Sherpa volunteers and to J. Callaghan, K. Hart, B. Pauly, and A. Mokashi for their assistance.

The work was supported in part by National Institutes of Health Grants HL-26533, HL-24335, and HL-08899.

REFERENCES

1. ANDERSON, B., R. A. KENNEY, AND E. NEIL. The role of chemoreceptors of the carotid and aortic regions in the production of the Mayer Waves. *Acta Physiol. Scand.* 20: 203–220, 1950.
2. BAND, D. M., P. WILSHAW, AND C. B. WOLFF. The speed of response of the carotid body chemoreceptor. In: *Morphology and Mechanisms of Chemoreceptors*, edited by A. S. Paintal. Delhi, India: Univ. Delhi Press, 1976, p. 197–207.
3. BARNARD, P., R. ZHANG, N. SMATRESK, M. POKORSKI, A. MOKASHI, AND S. LAHIRI. Carotid chemoreceptor and ventilatory responses in chronically hypoxic cats (abstr.). *Physiologist* 24: 114, 1981.
4. BLACK, A. M. S., D. I. McCLOSKEY, AND R. W. TORRANCE. The responses of carotid body chemoreceptors in the cat to sudden changes in hypercapnic and hypoxic stimuli. *Respir. Physiol.* 13: 36–49, 1971.
5. BLACK, A. M. S., AND R. W. TORRANCE. Respiratory oscillations in chemoreceptor discharge in the control of breathing. *Respir. Physiol.* 13: 221–237, 1971.
6. BLAYO, M. C., J. P. MACVERNES, AND J. J. POCIDALO. pH, P_{CO_2} and P_{O_2} of cisternal cerebrospinal fluid in high altitude natives. *Respir. Physiol.* 19: 298–311, 1973.
7. BROWN, H. W., AND F. PLUM. The neurological basis of Cheyne-Stokes respiration. *Am. J. Physiol.* 30: 849–861, 1961.
8. BRUSIL, P. J., T. B. WAGGENER, R. E. KRANAUER, AND P. GULESIAN. JR. Methods for identifying respiratory oscillations disclose altitude effects. *J. Appl. Physiol.: Respirat. Environ. Exercise Physiol.* 48: 545–556, 1980.
9. BULOW, K. Respiration and wakefulness in man. *Acta Physiol. Scand. Suppl.* 59: 1–110, 1963.
10. CHERNIACK, N. S. Respiratory dysrhythmias during sleep. *N. Engl. J. Med.* 305: 325–330, 1981.
11. CHERNIACK, N. S., C. VON EULER, I. HOMMA, AND F. F. KAO. Experimentally induced Cheyne-Stokes breathing. *Respir. Physiol.* 37: 185–200, 1979.
12. CHEYNE, J. A case of apoplexy, in which the fleshy part of the heart was converted into fat. *Dublin Hosp. Rep.* 2: 216, 1818.
13. DE CASTRO, F. Sur la structure et l'innervation du sinus carotidien de l'homme et des mammiférès. Nouveaux faits sur l'innervation et du fonction du glomus caroticum. Études anatomiques et physiologiques. *Trab. Lab. Invest. Biol. Univ. Madrid* 25: 331–380, 1928.
14. DEJOURS, P. Chemoreflexes in breathing. *Physiol. Rev.* 42: 335–358, 1962.
15. DEMPSEY, J. A., AND H. V. FORSTER. Mediation of ventilatory adaptations. *Physiol. Rev.* 62: 262–346, 1982.
16. DOUGLAS, C. G., AND J. S. HALDANE. The causes of periodic or Cheyne-Stokes breathing. *J. Physiol. London* 38: 401–419, 1909.
17. DOUGLAS, C. G., J. S. HALDANE, Y. HENDERSON, AND E. C. SCHNEIDER. Physiological observations made on Pike's Peak, Colorado, with special reference to adaptation to low barometric pressures. *Philos. Trans. R. Soc. London Ser. B* 203: 185–381, 1913.
18. DOUGLAS, N. J., D. P. WHITE, J. V. WEIL, C. L. PICKETT, R. J. MARTIN, D. W. HUDGEL, AND C. W. ZWILLICH. Hypoxic ventilatory response decreases during sleep in normal men. *Am. Rev. Respir. Dis.* 125: 286–289, 1982.
19. EDELMAN, N. H., P. E. EPSTEIN, S. LAHIRI, AND N. S. CHERNIACK. Ventilatory responses to transient hypoxia and hypercapnia in man. *Respir. Physiol.* 17: 302–314, 1973.
20. ELDRIDGE, F. L. Subthreshold central neural respiratory activity and after-discharge. *Respir. Physiol.* 39: 327–343, 1980.
21. FENNER, A., U. SCHALK, H. HOENICKE, A. WENDENBURG, AND T. ROCHLING. Periodic breathing in premature and neonatal babies: incidence, breathing pattern, respiratory gas tensions, response to changes in the composition of ambient air. *Pediatr. Res.* 7: 174–183, 1973.
22. GUILLEMINAULT, C. J., AND W. C. DEMENT (editors). *Sleep Apnea Syndromes.* New York: Liss, 1978.

23. GUYTON, A. C., J. W. CROWELL, AND J. W. MOORE. Basic oscillating mechanism of Cheyne-Stokes breathing. *Am. J. Physiol.* 187: 395–398, 1956.
24. HEYMANS, C., J. J. BOUCKAERT, AND L. DAUTREBANDE. Sinus carotidien et reflexes respiratoirs. II. Influences respiratoires reflexes de l'acidose, de l'alcalose, de l'anhydride carbonique, de l'ion hydrogene et de l'anoxemie: sinus carotidiens et les changes respiratoirs dans les poumons et au des poumons. *Arch. Int. Pharmacodyn. Ther.* 39: 400–408, 1930.
25. HURTADO, A. Animals in high altitudes: resident man. In: *Handbook of Physiology. Adaptation to the Environment*, edited by D. B. Dill and E. F. Adolf. Washington, DC: Am. Physiol. Soc., 1964, sect. 4, chapt. 54, p. 843–860.
26. KELLOGG, R. H. Some high points in high altitude physiology. In: *Environmental Stress*, edited by L. J. Folinsbee, J. A. Wagner, J. F. Borgia, B. L. Drinkwater, J. A. Gilver, and J. F. Bedi. New York: Academic, 1978, p. 317–323.
27. KHOO, M. C. K., R. E. KRONAUER, K. P. STROHL, AND A. S. SLUTSKY. Factors including periodic breathing in humans: a general model. *J. Appl. Physiol.: Respirat. Environ. Exercise Physiol.* 53: 644–659, 1982.
28. LAHIRI, S. Physiological response and adaptations to high altitude. In: *Environmental Physiology II*, edited by D. Robertshaw. Baltimore, MD: University Park, 1977, vol. 15, p. 217–251. (Int. Rev. Physiol. Ser.)
29. LAHIRI, S. Adaptive respiratory regulation—lessons from high altitude. In: *Environmental Physiology: Aging, Heat and Altitude*, edited by S. J. Horvath and M. K. Yousef. New York: Elsevier, 1981, p. 341–350.
30. LAHIRI, S., AND J. S. MILLEDGE. Acid-bases in Sherpa altitude residents and lowlanders at 4880m. *Respir. Physiol.* 2: 323–334, 1967.
31. LAHIRI, S., K. MARET, AND M. G. SHERPA. Dependence of high-altitude sleep apnea on ventilatory sensitivity to hypoxia. *Respir. Physiol.* 52: 281–301, 1983.
32. LAHIRI, S., A. MOKASHI, R. G. DELANEY, AND A. P. FISHMAN. Arterial P_{O_2} and P_{CO_2} stimulus threshold for carotid chemoreceptors and breathing. *Respir. Physiol.* 34: 359–375, 1978.
33. LAHIRI, S., E. MULLIGAN, AND A. MOKASHI. Adaptive responses of carotid body chemoreceptors to CO_2. *Brain Res.* 234: 137–147, 1982.
34. LAHIRI, S., N. J. SMATRESK, AND E. MULLIGAN. Responses of peripheral chemoreceptors to natural stimuli. In: *Physiology of Peripheral Arterial Chemoreceptors*, edited by H. Acker and R. O'Regan. Amsterdam: Elsevier, 1983, p. 221–256.
35. LANGE, R. L., AND H. H. HECHT. The mechanism of Cheyne-Stokes respiration. *J. Clin. Invest.* 41: 42–52, 1962.
36. LEFRANCOIS, R., H. GAUTIER, AND P. PASQUIS. Ventilatory oxygen drive in acute and chronic hypoxia. *Respir. Physiol.* 4: 217–228, 1968.
37. MICHEL, C. C., AND J. S. MILLEDGE. Respiratory regulation in man during acclimatization to high altitude. *J. Physiol. London* 168: 631–643, 1963.
38. MILLEDGE, J. S., AND S. LAHIRI. Respiratory control in lowlanders and Sherpa highlanders at altitude. *Respir. Physiol.* 2: 310–322, 1967.
39. MOSSO, A. La respiration periodique et la respiration superflue ou de luxe. *Arch. Ital. Biol.* 7: 48–127, 1886.
40. PAPPENHEIMER, J. R. Sleep and respiration of rats during hypoxia. *J. Physiol. London* 266: 191–207, 1977.
41. PHILLIPSON, E. A. Control of breathing during sleep. *Am. Rev. Respir. Dis.* 118: 768–774, 1975.
42. PREISS, G., S. ISCOE, AND C. POLOSA. Analysis of a periodic breathing pattern associated with Mayer waves. *Am. J. Physiol.* 228: 768–774, 1975.
43. PRYOR, W. W. Cheyne-Stokes respiration in patients with cardiac enlargement and prolonged circulation time. *Circulation* 4: 233–238, 1951.
44. READ, D. J. C. A clinical method for assessing the ventilatory response to carbon dioxide. *Australas. Ann. Med.* 16: 20–32, 1967.
45. REED, D. J., AND R. H. KELLOGG. Changes in respiratory response to CO_2 during natural sleep at sea level and at altitude. *J. Appl. Physiol.* 13: 325–330, 1958.
46. SCHLAFKE, M. E. Central chemosensitivity: a respiratory drive. *Rev. Physiol. Biochem. Pharmacol.* 90: 171–244, 1981.
47. SEVERINGHAUS, J. W., C. R. BAINTON, AND A. CARCELÉN. Respiratory insensitivity to hypoxia in chronically hypoxic man. *Respir. Physiol.* 1: 308–334, 1966.
48. SEVERINGHAUS, J. W., AND A. CARCELÉN B. Cerebrospinal fluid in man native to high altitude. *J. Appl. Physiol.* 19: 319–321, 1964.
49. SMATRESK, N. J., M. POKORSKI, AND S. LAHIRI. Opposing effects of dopamine receptor blockade on ventilation and carotid chemoreceptor activity. *J. Appl. Physiol.: Respirat. Environ. Exercise Physiol.* 54: 1567–1573, 1983.
50. STOKES, W. *The Diseases of the Heart and Aorta.* Dublin, Ireland: Hodges & Smith, 1854, p. 302–337.
51. STRANGE-PETERSEN, E., B. J. WHIPP, D. G. DRYSDALE, AND D. J. CUNNINGHAM. Carotid arterial blood gas oscillations and the phase of the respiratory cycle during exercise in man: testing a model. In: *The Regulation of Respiration During Sleep and Anesthesia*, edited by R. S. Fitzgerald, H. Gautier, and S. Lahiri. New York: Plenum, 1978, p. 335–342.
52. WEIL, J. V., E. BYRNE-QUINN, E. INGVAR, G. F. FILLY, AND R. F. GROVER. Acquired attenuation of chemoreceptor function in chronically hypoxic man at high altitude. *J. Clin. Invest.* 50: 186–195, 1971.
53. WEIL, J. V., M. H. KRYGER, AND C. H. SCOGGIN. Sleep and breathing at high altitude. In: *Sleep Apnea Syndromes*, edited by C. J. Guilleminault and W. C. Dement. New York: Liss, 1978, p. 119–136.
54. WEST, J. B. (editor). *High Altitude Physiology.* Stroudsburg, PA: Hutchinson, 1981, p. 363–368.

8

Ventilatory Control During Sleep in Normal Humans

JOHN V. WEIL, DAVID P. WHITE, NEIL J. DOUGLAS, AND CLIFFORD W. ZWILLICH

Cardiovascular Pulmonary Research Laboratory, University of Colorado Health Sciences Center, Denver, Colorado

SOME ASPECTS OF VENTILATORY CONTROL in normal humans during sleep are reviewed in this chapter as a standard against which alterations in sleep and sleep-induced changes in breathing at high altitude can be understood.

Although Dickens (8) described the index case of massive obesity with severe daytime somnolence, only recently has this syndrome been recognized as an expression of marked sleep disturbance caused by obstructive sleep apnea. The studies by Gastaut and others (14) in the mid 1960s awakened many of us to the concept that sleep can lead to disturbed breathing and that disturbed breathing can lead to poor quality sleep with major deleterious effects on the quality of life during wakefulness. These early studies soon triggered major growth in research concerned with sleep-induced alterations in respiratory control, which has been excellently reviewed (6, 17, 23). Much of the information derived from this work is consistent with the general concept that responses to a variety of respiratory stimuli (e.g., hypoxia, hypercapnia, or resistive loading on ventilatory effort), which are quite marked during wakefulness, are attenuated by sleep (Fig. 1; 3, 24–26, 28). This attenuation is usually more pronounced in rapid-eye-movement (REM) sleep, i.e., dreaming or paradoxical sleep, than in non-rapid-eye-movement (non-REM) sleep, i.e., quiet sleep. Sleep may largely abolish responses to other stimuli (such as the cough response to airway irritation), and only when such stimulation leads to arousal does the respiratory effect (cough) result (30).

Most information on sleep-induced alterations in ventilatory control has been derived from studies in patients and experimental animals. With the important exception of the classic work of Bulow (5), there have been relatively few studies of these phenomena in normal human subjects in the past. Recently, however, work by several laboratories, including our own, has focused

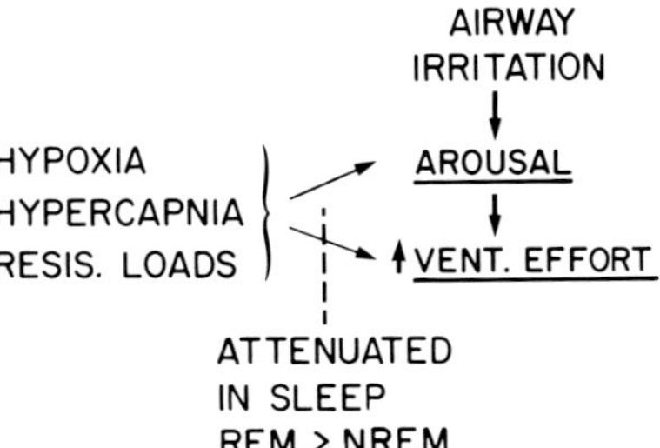

FIG. 1. Schematic representation of effects of sleep on ventilatory drives. REM, rapid eye movement; NREM, non rapid eye movement.

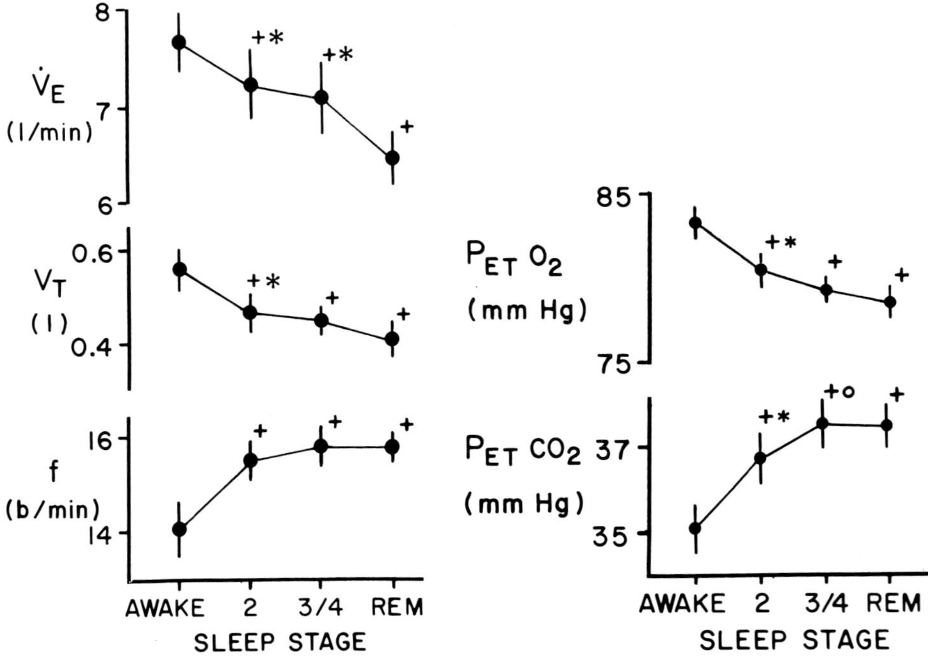

FIG. 2. Influence of sleep on ventilation, ventilatory pattern, and the partial pressure of end-tidal gases (PET_{O_2} and PET_{CO_2}) in normal subjects. At all stages of sleep, ventilation decreases and a rapid, shallow breathing pattern occurs. These changes are most pronounced in REM sleep. Hypercapnia and hypoxia occur in all sleep stages also, suggesting a reduced alveolar ventilation. †$P < 0.05$ vs. awake; *$P < 0.05$ vs. REM. $\dot{V}_E$, expired ventilation; V_T, tidal volume; f, respiratory frequency. [Data from 10.]

on ventilatory control during sleep in normal humans, providing a standard for judging findings in patients. This work is the major theme of this chapter.

VENTILATION DURING SLEEP

Studies of ventilation and ventilatory pattern during sleep in 19 healthy adult subjects showed that ventilation was significantly reduced in all sleep stages compared to wakefulness (Fig. 2; 10). Ventilation was most reduced in

REM sleep, where it averaged 84% of waking values. In all sleep stages ventilatory pattern shifted toward more rapid, shallow breathing, changes that were also most prominent in REM sleep. Inspiratory flow rates remained at waking values in non-REM but were reduced by ~20% in REM sleep. No significant changes were seen in inspiratory duty cycle (inspiratory time/total time). End-tidal gas tensions reflected the combined effects of decreased total ventilation and decreased efficiency of the tachypneic breathing pattern in that end-tidal O_2 tensions were decreased and CO_2 tensions rose comparably (Fig. 2).

We have no clear explanation for the absence of an increase in CO_2 tension between non-REM and REM sleep despite reductions of ventilation and tidal volume. However, the increased variability of tidal volume and hence end-tidal CO_2 tension typical of REM sleep might have obscured such changes. Overall our findings are similar to those reported by others (5, 13, 15, 27, 31), although in one study no decrease in ventilation was found in quiet sleep (13).

HYPERCAPNIC VENTILATORY RESPONSE DURING SLEEP

That the ventilatory response to hypercapnia is depressed in sleep is well known (5, 27), but the relationship of such changes to specific sleep stages, particularly REM sleep, in normal man has not been clear. Accordingly we measured the ventilatory response to hypercapnia by using hyperoxic rebreathing techniques in 12 normal subjects (12). Hypercapnic ventilatory responses were clearly diminished in all stages of sleep (Fig. 3), averaging less than one-half of the waking value in non-REM and less than one-third of the waking level during REM sleep. In addition to the decreases in slope of the hypercapnic response, sleep induced impressive and progressive rightward shifts in the position of these responses, i.e., to higher CO_2 tensions. Thus during sleep CO_2 tension would have to rise considerably more than during wakefulness before hypercapnia significantly stimulates ventilation. In REM sleep, for example, this increase in CO_2 tension would have to be as great as 6 or 7 mmHg above waking values before ventilatory stimulation could be expected. This shift of the hypercapnic response in sleep may play an important permissive role in

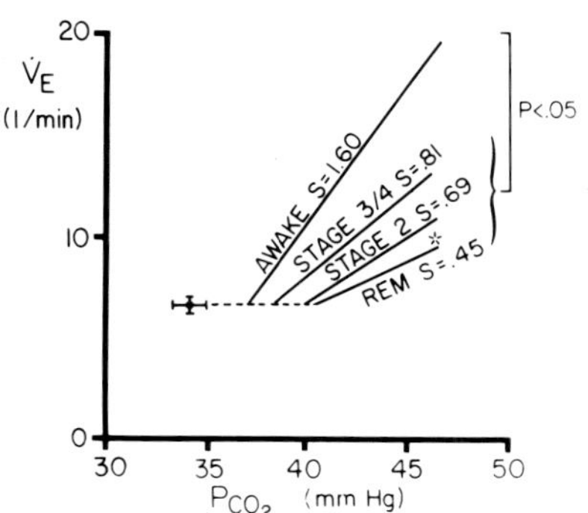

FIG. 3. Influence of sleep on hypercapnic ventilatory response in a normal man. Sleep induces depression of CO_2 tension (P_{CO_2}) slope and a rightward shift in response compared to waking values. Both effects are greatest in REM sleep. [Data from 12.]

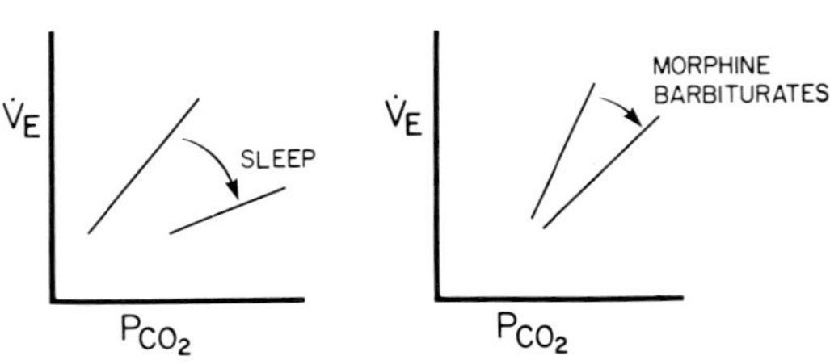

FIG. 4. Schematic representation of contrasting effects of sleep (*left panel*) and pharmacological ventilatory depressants (*right panel*) on ventilatory response to hypercapnia. In contrast to isolated effects of pharmacological depressants on slope, sleep both depresses the slope and shifts the curve to right. [Data from 12, 21, 34.]

the development of hypercapnia during sleep, i.e., it may explain the failure of sleep-associated hypercapnia to initiate corrective increases in ventilation. Because these shifts were largely measured over a nonphysiological, hypercapnic range, their relevance to the CO_2 tension–ventilation relationship at near-normal CO_2 tensions is not entirely certain. However, these shifts were generally associated with parallel changes in ventilation–CO_2 tension rest points (i.e., base-line values) during various sleep stages, suggesting that the observed shifts are probably relevant to physiological events. The shift in the CO_2 response also distinguishes the ventilatory depressant effects of sleep from those of pharmacological ventilatory depressants (e.g., barbiturates or morphine), which typically decrease the slope with little rightward shift (Fig. 4; 21, 34). In fact not many experimental interventions can produce an acute rightward shift in the CO_2 response analogous to that seen during sleep. Metabolic alkalosis can, but this does not seem particularly relevant to sleep. Cherniack and colleagues (7) showed that cooling the ventral medullary surface in animals produced changes remarkably similar to those seen in man during sleep and thus might provide some clues to the locus and mechanism of sleep-related effects on the CO_2 response.

HYPOXIC VENTILATORY RESPONSE DURING SLEEP

There are few studies of the hypoxic ventilatory response in normal or abnormal human subjects during sleep. Work in animals suggested that the hypoxic response may be better maintained than the hypercapnic response during REM sleep (26). We therefore measured the ventilatory response to isocapnic hypoxia in six normal men and found that the slope of this response behaved similar to that for the hypercapnic response; it was reduced in all sleep stages compared to wakefulness with a marked reduction to less than one-third of awake values in REM sleep (Fig. 5; 11). Thus these findings differ from those in the sleeping dog, which showed that the hypoxic response was maintained at waking levels in non-REM and REM sleep in contrast to the clear sleep-associated decrements for the hypercapnic response. Our findings also differ from those of Gothe et al. (16), who found no decrease in hypoxic response during sleep in men. However, in that study there was no attempt to maintain isocapnic conditions, and the resulting hypocapnic alkalosis probably depressed the hypoxic response, which reduced the chances of finding a further

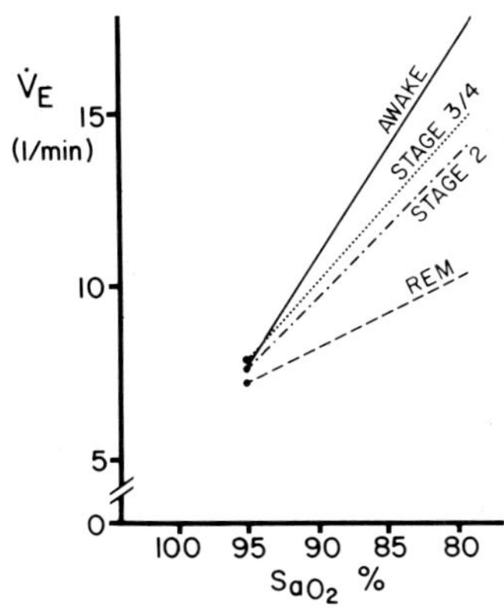

FIG. 5. Influence of sleep on ventilatory response to isocapnic hypoxia in 6 normal men. Hypoxic response was significantly decreased in all sleep stages in comparison to wakefulness; the most marked changes were in REM sleep. Sa_{O_2}, arterial O_2 saturation. [Data from 11.]

depression in sleep. Hedemark and Kronenberg (19) also found no consistent decrease in hypoxic response in non-REM sleep, but their study group included both men and women, which for reasons discussed in the next section may have influenced their results. Our data agree with those of Berthon-Jones and Sullivan (1), who made isocapnic hypoxic measurements comparable to ours.

Thus, in normal man, quiet or non-REM sleep is associated with decreased total ventilation, a rapid, shallow breathing pattern, mild hypoxia and hypercapnia, and moderate decrements in hypercapnic and hypoxic ventilatory responses. These changes are more pronounced in REM sleep with further decreases in total ventilation, tidal volume, and inspiratory flow rate and with major reductions in ventilatory responses to both hypoxia and hypercapnia. Although we have focused on sleep-related decreases in ventilation, changes in ventilation in these normal subjects are small. In other words, ventilation is relatively well maintained in the sleep of normal subjects despite large decrements in ventilatory drives. Indeed the question of whether ventilation in sleep is related to hypoxic or hypercapnic ventilatory drives remains unanswered. Under normal circumstances these drives seem to make a relatively minor contribution to ventilation, but they may be more important in patients in whom alteration of these drives may be pivotal in the development of hypoxemia and hypercapnia, particularly during REM sleep.

SEX DIFFERENCES IN VENTILATORY CONTROL DURING WAKEFULNESS AND SLEEP

Because apnea, hypoventilation, and hypoxemia in sleep are far more common in men than in women (2), normal subjects of both sexes have been studied to determine whether there are male-female differences in ventilatory control during sleep. No such differences were found for the hypercapnic response (12). Men and women showed the moderate decrease of the CO_2 response in quiet sleep and further decrement in REM sleep that had been seen earlier in men (Fig. 6). In contrast there were impressive differences in

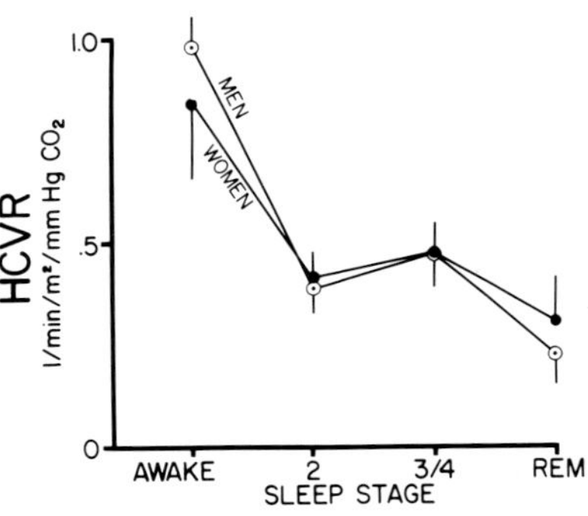

FIG. 6. Comparison of influence of sleep on slope of hypercapnic ventilatory response (HCVR) in men ($n = 6$) and women ($n = 6$). Sleep had comparable effects in the 2 sexes (12).

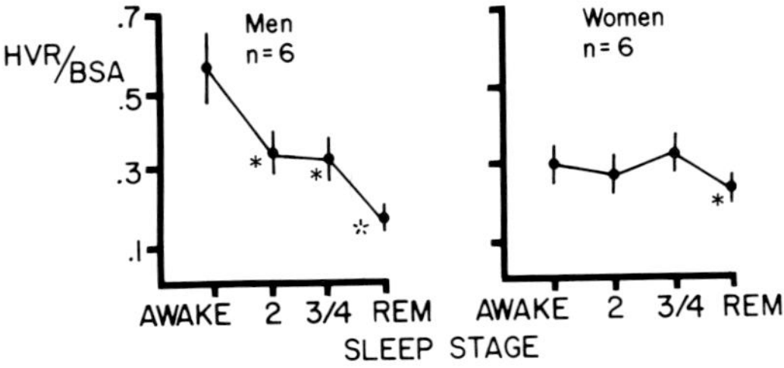

FIG. 7. Influence of sleep on the hypoxic ventilatory response (HVR) in men and women. In men, sleep induced stage-dependent decrements in HVR. Although women started from a lower waking value, their hypoxic response was better preserved during all sleep stages, with only a slight decrement noted in REM (*$P < 0.05$) (11, 35). BSA, body surface area.

the effect of sleep on the hypoxic ventilatory response; the women, starting at a lower awake value, showed much better preservation of the hypoxic response during sleep than men (Fig. 7; 35). The lower waking hypoxic responses raised the question of whether the absence of sleep-related depression of hypoxic ventilatory response was merely a function of a lower base-line (awake) value. However, comparison of a subgroup of men with relatively low waking hypoxic responses with a subgroup of women with relatively high awake values (which permitted men and women with similar starting values to be compared) showed that the differences remained, i.e., much greater stage-related depression of hypoxic response in men than in women.

The lower hypoxic response during wakefulness in women compared to men was surprising because progesterone, a hormone typically present in premenopausal women, is a ventilatory stimulant that augments the hypoxic ventilatory response (38). We found no systematic study comparing waking ventilatory responses in men and women, although fragmentary data suggested that normal women may indeed have lower ventilatory responses to chemical stimuli than men (21, 29). Consequently we compared hypoxic responses in 10 normal menstruating females and 12 normal men (36). The hypoxic ventilatory response, measured as the slope parameter A, was significantly lower ($P < 0.05$) in women, 109 ± 13 (SEM), than in men, who averaged 167 $\pm$ 22. However, as others have seen (29), there was a moderate increase of ~40% in the women's response during the luteal phase over that in the follicular phase of the menstrual cycle. Comparable differences in the hypercapnic response were also observed between the sexes and between the luteal and follicular phases of the menstrual cycle.

Interpreting male-female differences was confounded by differences in body size: the women were smaller than the men. To determine whether an index of body size should be used to correct for these differences, we examined correlations of hypoxic ventilatory response with several size-related variables for both men and women; this led to a surprising finding. In men hypoxic response was not significantly related to height, weight, or body surface area but was highly correlated with O_2 consumption and CO_2 production. However, in women the hypoxic response was related to none of these variables. In men the strongest correlation was that of hypoxic ventilatory response to the relative metabolic rate, i.e., O_2 consumption or CO_2 production corrected for body surface area. Again, however, no such relationship was found in the women, which particularly intrigued us. Several previous studies had shown a strong relationship between metabolic rate and the hypoxic response in a wide variety of hypermetabolic states including exercise (33), hyperthermia (22), postprandial state (40) (where both are increased) and in hypometabolic states such as myxedema (39) and semistarvation (9) (where the hypoxic response is decreased). It is intriguing that virtually all of this work has been done in men. At any rate hypoxic ventilatory response seems closely linked to metabolic rate in men but not in women. Although the reasons for this difference are unclear, it may reflect a fundamental difference in ventilatory control between the sexes.

METABOLIC RATE DURING SLEEP

An effect of metabolic rate on hypoxic ventilatory response may be relevant to ventilatory control during sleep, because, as has long been recognized, metabolic rate decreases in sleep (4, 27). Our studies confirm this, showing that in most individuals metabolic rate decreases progressively with increasing sleep duration, reaching a nadir sometime in the early morning (37). Subsequently it may gradually increase or remain depressed until awakening, when there is an abrupt increase to a level comparable to that which existed prior to sleep. The pattern of metabolic rate depression appeared to be largely a function of the preceding duration of sleep with no significant variation as a function of sleep stage, although others have reported small

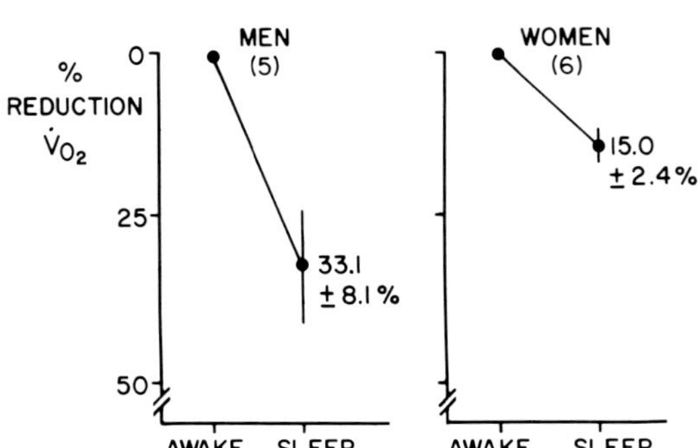

FIG. 8. Maximal decrements in metabolic rate during sleep in men and women. Reductions in O_2 consumption ($\dot{V}O_2$) during sleep were more consistent and on the average 2 times as great in men than in women ($P < 0.05$) (37).

increases of ~5% in metabolic rate asssociated with REM sleep (4, 18). A particularly intriguing feature of our findings was the variation between individuals in the extent to which metabolic rate decreased during sleep, ranging from almost no decrease to maximal decreases as high as 50% of awake values. Sex differences for this effect are worth noting: the smallest decreases were found almost exclusively in women, who averaged a 15% decrement in metabolic rate, whereas the larger decreases (on occasion as great as 50%) were seen in the men, who showed an average decrease of 33% (Fig. 8). The three men with the greatest decrements of metabolic rate (decreases of ~50%) were the only subjects in whom erratic breathing occurred, which coincided with the maximum depression of metabolic rate in two of them. We also found that in men the sleep-associated decrement in metabolic rate seemed to clearly increase with age, as Webb and Heistand (32) had observed, but no such effect was evident in women (Fig. 9).

Thus hypercapnic ventilatory responses decrease during sleep similarly in men and women. However, whereas the hypoxic ventilatory response is decreased during sleep in men, it is maintained at near waking levels in women although they start from lower absolute values. In men the hypoxic ventilatory response is strongly influenced by changes in metabolic rate, whereas in women no such relationship is evident. The decreased metabolic rate associated with sleep is substantially greater in men than in women, which might partially account for the decreased hypoxic ventilatory response seen in men during sleep. Conversely women may retain a better hypoxic ventilatory response in sleep because they tend to preserve their sleeping metabolic rate at near the wakefulness level and because changes in metabolic rate may influence their hypoxic ventilatory response less (Fig. 10). However, it is also clear that, although hypoxic ventilatory response may be linked to metabolic rate, there are also decreases in hypoxic response related to REM sleep, especially in men, that cannot be accounted for by metabolic rate.

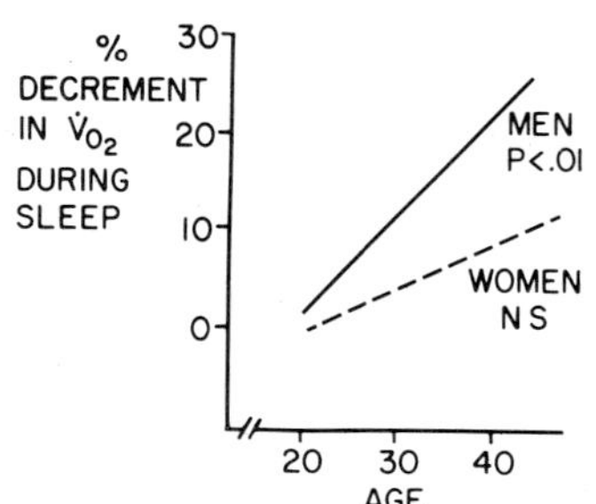

FIG. 9. Sleep-related decrements in metabolic rate as a function of age. Extent to which $\dot{V}O_2$ fell during sleep was greater with increasing age in men; no such relationship was evident in women (37).

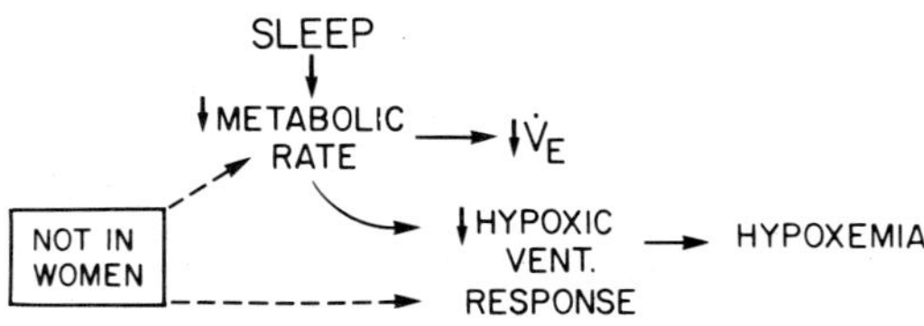

FIG. 10. Schematic representation of potential interrelationships between sleep, metabolic rate, ventilation, and hypoxic ventilatory response. Hypoventilation and hypoxemia may be less common in women partly because women show better preservation of metabolic rate and hypoxic ventilatory response in sleep.

SUMMARY

In normal humans, sleep causes decreased ventilation with a rapid, shallow breathing pattern, which is especially pronounced in REM sleep. There are stage-dependent decrements in the ventilatory response to hypercapnia that consist of a right shift in curve position and a decreased slope, with the greatest changes in REM sleep. Similarly the greatest decrements in slope of the hypoxic response also occurred in REM sleep. Hypercapnic responses are altered similarly in men and women during sleep, whereas the hypoxic response, although starting from a lower value, seems much better preserved during sleep in women than in men. Some of this resistance to the depression of the hypoxic response in sleep in women may be related to the findings that their hypoxic response appears to be less sensitive to changes in metabolic rate and that the decrement in metabolic rate with sleep appears to be much smaller than in men.

REFERENCES

1. BERTHON-JONES, M., AND C. E. SULLIVAN. Ventilatory and arousal responses to hypoxia in sleeping humans. *Am. Rev. Respir. Dis.* 125: 632–639, 1982.
2. BLOCK, A. J., P. G. BOYSEN, J. W. WYNNE, AND L. A. HUNT. Sleep apnea, hypopnea and oxygen desaturation in normal subjects. A strong male predominance. *N. Engl. J. Med.* 300: 513–517, 1979.
3. BOWES, G., E. R. TOWNSEND, S. M. BROMLEY, L. F. KOZAR, AND E. A. PHILLIPSON. Role of the carotid body and of afferent vagal stimuli in the arousal response to airway occlusion in sleeping dogs. *Am. Rev. Respir. Dis.* 123: 644–647, 1981.
4. BREBBIA, D. R., AND K. Z. AKLTSHULER. Oxygen consumption rate and electroencephalographic stage of sleep. *Science* 150: 1621–1623, 1965.
5. BULOW, K. Respiration and wakefulness in man. *Acta Physiol. Scand. Suppl.* 209: 1963.
6. CHERNIACK, N. S. Respiratory dysrhythmias during sleep. *N. Engl. J. Med.* 305: 325–330, 1981.
7. CHERNIACK, N. S., C. VON EULER, I. HOMMA, AND F. F. KAO. Graded changes in central chemoceptor input by local temperature changes on the ventral surface of medulla. *J. Physiol. London* 287: 191–211, 1979.
8. DICKENS, C. *The Posthumous Papers of the Pickwick Club.* London: Chapman & Hall, 1837.
9. DOEKEL, R. C., C. W. ZWILLICH, J. O. BATTAGLIA, E. K. COTTON, AND J. V. WEIL. Clinical semi-starvation: depression of hypoxic ventilatory response. *N. Engl. J. Med.* 295: 358–361, 1976.
10. DOUGLAS, N. J., D. P. WHITE, C. K. PICKETT, J. V. WEIL, AND C. W. ZWILLICH. Respiration during sleep in normal man. *Thorax* 37: 840–844, 1982.
11. DOUGLAS, N. J., D. P. WHITE, J. V. WEIL, C. K. PICKETT, R. J. MARTIN, D. W. HUDGEL, AND C. W. ZWILLICH. Hypoxic ventilatory response decreases during sleep in normal men. *Am. Rev. Respir. Dis.* 125: 286–289, 1982.
12. DOUGLAS, N. J., D. P. WHITE, J. V. WEIL, C. K. PICKETT, AND C. W. ZWILLICH. Hypercapnic ventilatory response in sleeping adults. *Am. Rev. Respir. Dis.* 126: 758–762, 1982.
13. DURON, B. Respiratory function during physiological sleep. *Bull. Physiopath. Respir.* 8: 1031–1057, 1972.
14. GASTAUT, H., C. A. TASSINARI, AND B. DURON. Polygraphic study of episodic diurnal and nocturnal (hypnic and respiratory) manifestations of the Pickwick syndrome. *Brain Res.* 2: 167–186, 1966.
15. GOTHE, B., M. D. ALTOSE, M. D. GOLDMAN, AND N. S. CHERNIACK. Effect of quiet sleep on resting and CO_2-stimulated breathing in humans. *J. Appl. Physiol.: Respirat. Environ. Exercise Physiol.* 50: 724–730, 1981.
16. GOTHE, B., M. D. GOLDMAN, N. S. CHERNIACK, AND P. MANTEY. Effect of progressive hypoxia on breathing during sleep. *Am. Rev. Respir. Dis.* 126: 97–102, 1982.
17. GUILLEMINAULT, C., AND W. C. DEMENT (editors). *Sleep Apnea Syndromes.* New York: Liss, 1978.
18. HASKELL, E. H., J. W. PALCA, J. M. WALKER, R. J. BERGER, AND H. C. HELLER. Metabolism and thermoregulation during stages of sleep in humans exposed to heat and cold. *J. Appl. Physiol.: Respirat. Environ. Exercise Physiol.* 51: 948–954, 1981.
19. HEDEMARK, L. L., AND R. S. KRONENBERG. Ventilatory and heart rate responses to hypoxia and hypercapnia during sleep in adults. *J. Appl. Physiol.: Respirat. Environ. Exercise Physiol.* 53: 307–312, 1982.
20. HIRSHMAN, C. A., R. E. McCULLOUGH, P. J. COHEN, AND J. V. WEIL. Hypoxic ventilatory drive in dogs during thipental, ketamine, or pentobarbital anesthesia. *Anesthesiology* 43: 628–634, 1975.
21. HIRSHMAN, C. A., R. E. McCULLOUGH, AND J. V. WEIL. Normal values for hypoxic and hypercapnic ventilatory drives in man. *J. Appl. Physiol.* 38: 1095–1098, 1975.
22. NATALINO, M. R., C. W. ZWILLICH, AND J. V. WEIL. Effects of hyperthermia on control of ventilation. *J. Lab. Clin. Med.* 89: 564–572, 1977.
23. PHILLIPSON, E. A. Control of breathing during sleep. *Am. Rev. Respir. Dis.* 118: 909–939, 1978.
24. PHILLIPSON, E. A., L. F. KOZAR, AND E. MURPHY. Respiratory load compensation in awake and sleeping dogs. *J. Appl. Physiol.* 40: 895–902, 1976.
25. PHILLIPSON, E. A., L. F. KOZAR, A. S. REBUCK, AND

E. MURPHY. Ventilatory and waking responses to CO_2 in sleeping dogs. *Am. Rev. Respir. Dis.* 115: 251–259, 1977.

26. PHILLIPSON, E. A., C. E. SULLIVAN, D. J. C. READ, E. MURPHY, AND L. F. KOZAR. Ventilatory and waking responses to hypoxia in sleeping dogs. *J. Appl. Physiol.: Respirat. Environ. Exercise Physiol.* 44: 512–520, 1978.
27. ROBIN, E. D., R. D. WHALEY, C. H. CRUMP, AND D. M. TRAVIS. Alveolar gas tensions, pulmonary ventilation and blood pH during physiologic sleep in normal subjects. *J. Clin. Invest.* 37: 981–989, 1958.
28. SANTIAGO, T. V., A. K. SINHA, AND N. H. EDELMAN. Respiratory flow-resistive load compensation during sleep. *Am. Rev. Respir. Dis.* 123: 382–387, 1981.
29. SCHOENE, R., H. ROBERTSON, D. PIERSON, AND A. PETERSON. Respiratory drives and exercise in menstrual cycles of athletic and nonathletic women. *J. Appl. Physiol.: Respirat. Environ. Exercise Physiol.* 50: 1300–1305, 1981.
30. SULLIVAN, C. E., L. F. KOZAR, E. MURPHY, AND E. A. PHILLIPSON. Arousal, ventilatory, and airway responses to bronchopulmonary stimulation in sleeping dogs. *J. Appl. Physiol.: Respirat. Environ. Exercise Physiol.* 47: 17–25, 1979.
31. TABACHNIK, E., N. L. MULLER, A. C. BRYAN, AND H. LEVISON. Changes in ventilation and chest wall mechanics during sleep in normal adolescents. *J. Appl. Physiol.: Respirat. Environ. Exercise Physiol.* 51: 557–564, 1981.
32. WEBB, P., AND M. HIESTAND. Sleep metabolism and age. *J. Appl. Physiol.* 38: 257–262, 1975.
33. WEIL, J. V., E. BYRNE-QUINN, I. E. SODAL, J. S. KLINE, R. E. McCULLOUGH, AND G. F. FILLEY. Augmentation of chemosensitivity during mild exercise in normal man. *J. Appl. Physiol.* 33: 813–819, 1972.
34. WEIL, J. V., R. E. McCULLOUGH, J. S. KLINE, AND I. E. SODAL. Diminished ventilatory response to hypoxia and hypercapnia after morphine in normal man. *N. Engl. J. Med.* 292: 1103–1106, 1975.
35. WHITE, D. P., N. J. DOUGLAS, C. K. PICKETT, J. V. WEIL, AND C. W. ZWILLICH. Hypoxic ventilatory response during sleep in normal premenopausal women. *Am. Rev. Respir. Dis.* 126: 530–533, 1982.
36. WHITE, D. P., N. J. DOUGLAS, C. K. PICKETT, J. V. WEIL, AND C. W. ZWILLICH. Sexual influence on the control of breathing. *J. Appl. Physiol.: Respirat. Environ. Exercise Physiol.* 54: 874–879, 1983.
37. WHITE, D. P., C. K. PICKETT, J. V. WEIL, AND C. W. ZWILLICH. The influence of metabolic rate on breathing during sleep (abstr.). *Am. Rev. Resp. Dis.* 127: 236, 1983.
38. ZWILLICH, C. W., M. R. NATALINO, F. D. SUTTON, AND J. V. WEIL. Effects of progesterone on chemosensitivity in normal men. *J. Lab. Clin. Med.* 92: 262–269, 1978.
39. ZWILLICH, C. W., D. J. PIERSON, F. D. HOFELDT, E. G. LUFKIN, AND J. V. WEIL. Ventilatory control in myxedema and hypothyroidism. *N. Engl. J. Med.* 292: 662–665, 1975.
40. ZWILLICH, C. W., S. A. SAHN, AND J. V. WEIL. Effects of hypermetabolism on ventilation and chemosensitivity. *J. Clin. Invest.* 60: 900–906, 1977.

9

Hypoxia and Brain Blood Flow

NORMAN H. EDELMAN, TEODORO V. SANTIAGO,
AND JUDITH A. NEUBAUER

Pulmonary Diseases Division, Department of Medicine, University of Medicine and Dentistry of New Jersey–Rutgers Medical School, New Brunswick, New Jersey

THE ASCENT TO HIGH ALTITUDE presents a variety of physiological and psychological obstacles. The most prominent is the decline in ambient O_2 tension (Po_2); much of this volume is devoted to consideration of the effects of the resultant hypoxemia on respiration and circulation. It seems likely, however, that acute responses and adaptations to hypoxia in the central nervous system also play key roles in determining man's ability to ascend to high altitudes. Because, at least in the short run, the O_2 supply to the brain at a given Po_2 of inspired gas depends largely on ventilation and brain perfusion, the brain blood flow response to hypoxia must be crucial in the integrated physiological responses considered in this chapter.

As respiratory physiologists our major purpose has been to elucidate the roles of brain hypoxia and brain blood flow in the control of breathing. We have, for example, shown substantial effects of brain hypoxia on ventilation and the ventilatory responses to chemical and neurochemical respiratory stimuli (3–5, 8). Accordingly this chapter focuses on the modification of brain blood flow by hypoxia with emphasis on how these responses alter ventilation. However, brain hypoxia and the blood flow response to this stimulus have nonventilatory effects that are probably also important during the ascent to high altitude. Higher cortical function, for example, is almost definitely modified. Whether this is a hindrance (impairment of critical judgment) or an aid (creation of euphoria necessary to complete the arduous task) the climbers themselves must decide.

Acute mountain sickness may also be related to brain hypoxia. This syndrome includes headache, nausea, and malaise, suggesting that it is caused by brain edema. Whether the edema is simply hydrostatic or involves alterations in capillary permeability due to hypoxia is unclear. The issue is important for the control of brain perfusion because if hydrostatic pressure is the major determinant, individuals with the most brisk increases in brain blood flow might be the most susceptible. On the other hand, if the degree of tissue hypoxia per se is most important, those with the least increase might be most susceptible. In either case an individual's brain blood flow response to hypoxia probably plays an important role in his susceptibility to acute mountain sickness. This hypothesis has not been tested.

GENERAL CHARACTERISTICS OF BRAIN VASCULAR CONTROL

To establish the proper perspective for discussing the effects of hypoxia on brain blood flow we have briefly outlined the major elements of brain vascular control. More detailed discussions are available in recent reviews (15, 17, 27). A simple scheme is presented in Figure 1.

In a process similar to ventilatory control, tight coupling accomplishes the critical function of matching blood flow to metabolism. Not only does whole brain blood flow change appropriately with whole brain metabolism, but coupling can be shown regionally. Discrete stimuli elicit concordant increases in metabolism and blood flow in those areas of the brain that mediate the responses. Because of the brisk response of the brain vasculature to hypoxia and hypercapnia, the effect of metabolism has been thought by many to be mediated by responses to one or both of these variables. However, no direct evidence supports this idea, and the initially most attractive of these factors—interstitial $[H^+]$—may not satisfy all experimental findings (see CHARACTERISTICS OF BRAIN BLOOD FLOW RESPONSE TO HYPOXIA, p. 103).

Other variables that influence the brain vasculature include vasoactive ions (e.g., K^+), amines, peptides, prostaglandins, an intrinsic response of arteriolar smooth muscle to stretch (i.e., the Bayliss effect, which may explain pressure-flow autoregulation), and activity of its autonomic nervous supply. In addition the variables that influence flow in all vascular beds (e.g., blood viscosity) also apply to the circulation of the brain. Because neurogenic control may be involved in the response to hypoxia, it requires further comment.

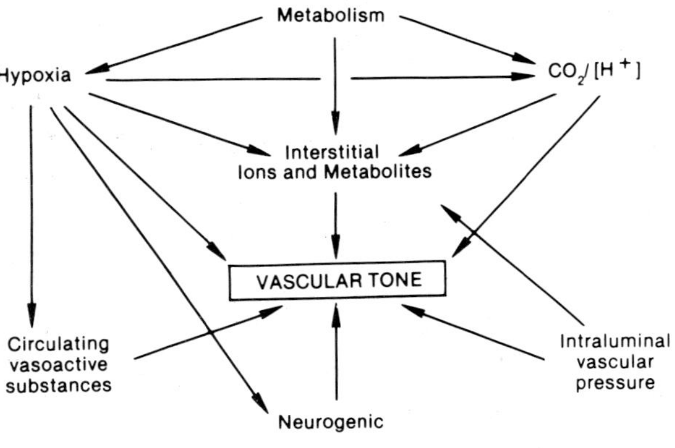

FIG. 1. Some proposed relationships of variables involved in regulating brain blood flow. Metabolism may alter vascular tone via hypoxia and acidosis or by altering interstitial ions and metabolites. Direct and indirect actions of $[H^+]$, hypoxia, and other factors have also been proposed.

After much disagreement, physiologically significant neural control of the brain circulation is now generally accepted, although there is still substantial controversy about the extent of the effect and reports of disconcertingly great species differences (10). Apparently one problem in demonstrating neural control stemmed from the exquisitely tight linkage of brain blood flow to metabolism, which counterbalances neural effects in the experimental setting. This dual-effects hypothesis, first advanced by Harper and co-workers (11), holds that neural control predominates in extraparenchymal vessels, whereas metabolic control resides in the small intraparenchymal vessels. Thus any change in flow by neurally induced vasomotion could be offset by opposite metabolically mediated vasomotion. This probably explains why adrenergic reduction of brain blood flow may be more easily demonstrated during maximum metabolic vasodilation caused by hypercapnia (11, 13). Not all data support the dual-effects concept (16).

Adrenergic and cholinergic receptors and responses have both been demonstrated. However, the physiological significance of adrenergic control is more firmly established than that of cholinergic control (9, 17).

Vascular innervation within the central nervous system has a nonhomogeneous distribution. The anterior, middle cerebral, internal carotid, and posterior communicating arteries all have more adrenergic nerve terminals than the posterior cerebral, basilar, and vertebral arteries (9). Therefore stimulating the extracerebral sympathetic inflow to the brain vasculature could cause nonuniform vasoconstriction with the greatest effect in the cortex (see *Regional Distribution*, p. 105).

CHARACTERISTICS OF BRAIN BLOOD FLOW RESPONSE TO HYPOXIA

Mediators

Many mediators for the brain vascular response to hypoxia have been proposed (Table 1; 2, 15, 17, 21), although none has proved wholly satisfactory. One attractive hypothesis has been an increase in interstitial $[H^+]$ (18) because it explains vascular responses to hypoxia, hypermetabolism, and change in

TABLE 1. *Potential mediators of brain vascular response to hypoxia*

	Pro	Con
Interstitial $[H^+]$	High sensitivity to CO_2 High potential for anaerobic glycolysis	Too slow Not always found
Interstitial $[K^+]$	Increases with hypoxia Appropriate dynamics	Quantitative problems
Adenosine	Potent vasodilator Released with increased neuronal activity	Not always found with hypoxia

$[Ca^{2+}]$ has been postulated as intermediate for some of above. Evidence is largely fragmentary for amines, peptides, prostaglandins, and reflexes.

acid-base status with one variable that readily dilates brain vessels and could be involved in the ventilatory response to chronic hypoxia. The high capacity of the brain for anaerobic glycolysis, the demonstration of anaerobic glycolysis during sea-level euoxia, and the ready increase in anaerobic glycolysis with relatively mild hypoxia all support the idea. The problem arises when the dynamic characteristics of the response are examined. Nilsson et al. (22) have shown that with step hypoxia, brain blood flow increases prior to a measurable increase in brain interstitial lactate concentration. Thus, although interstitial acidosis may contribute to the vasodilator response to brain hypoxia, other factors must also be involved.

As with other vascular beds, elaboration of K^+ and adenosine have been proposed to mediate the vascular response to both hypoxia and increased metabolism. However, quantitative problems arise with the suggestion of K^+ elaboration; that is, less is released during hypoxia than seems necessary for the vasodilation (1). Increases in adenosine levels are not always demonstrable during hypoxia (23). Others (34), however, find good correlation between hypoxia and adenosine levels in brain. It is probably most prudent now to conclude that the brain blood flow response to hypoxia reflects an interaction of more than one mechanism.

Physiological Stimulus

Clearly whatever receptor mechanism is involved in the brain vascular response to hypoxia, O_2 supply to the sensor is mostly limited by O_2 flow rather than by diffusion. That is, responses to lowered O_2 content of arterial blood (e.g., by hypoxic hypoxia, carboxyhemoglobinemia, or anemia) are generally equivalent whether or not accompanied by lowered Po_2 in arterial blood (8, 14, 30). That a leftward shift of the O_2-dissociation curve for hemoglobin increases brain blood flow indicates some diffusion limitation in the transport of O_2 to the sensing system (31).

These observations suggest that O_2 sensors that behave as if they are primarily diffusion limited, such as the carotid bodies, are not involved in the response. Two careful studies by Heistad et al. (12) and Traystman et al. (30) confirm that the cerebrovascular response to hypoxia may occur in the absence of the carotid bodies. The explanation of earlier findings, which led to the opposite conclusion, is not clear (24).

Dynamic Characteristics

The brain blood flow response to hypoxia is rapid but not instantaneous. We have done studies to quantitate the dynamic properties of this control system (7). Anesthetized and paralyzed goats were prepared so that brain blood flow could be measured continuously with an electromagnetic flow meter. Sinusoidal variations in the O_2 content of inspired gases, with variable amplitude and period, were produced with a specially constructed input device to the mechanical ventilator. To analyze the data, continuously recorded brain blood flow was related to continuously recorded O_2 saturation of arterial blood, Fourier analysis was performed, and Bode plots were constructed. The data showed that the mean level of the hypoxic stimulus seemed to have no major effect on the dynamic characteristics of the response. The response was adequately fit by a single exponential function with time constants varying from 30 to 40 s. This is rather long for a simple reflex and is consistent with the time required for diffusion of vasoactive substances from parenchyma to resistance vessels (32).

These findings may be important to breathing at high altitude because the dynamic characteristics of the cerebrovascular response to hypoxia probably influence those of the ventilatory response to hypoxia. In chapter 11 Cherniack and his co-workers show that cyclical variations in respiratory pattern (e.g., Cheyne-Stokes breathing, which is characteristic of lowlanders during sleep at high altitude) depend on predictable modifications of the gain of the ventilatory controller. However, the occurrence and nature of the periodic breathing also require and reflect the delay (or phase lag) between the site of gas exchange and the respiratory control sensor. Because brain blood flow modifies both P_{O_2} and P_{CO_2} in brain tissue and both alter ventilation, the relatively long time constant of the brain blood flow response to hypoxia is probably quantitatively significant in the generation of Cheyne-Stokes ventilation.

Regional Distribution

In the few published studies where it has been examined, no marked differences in the blood flow response to hypoxia were found between brain regions (12, 19). These studies were done in anesthetized animals. In one study on anesthetized rabbits, in which the increase in small-vessel blood volume was measured, a lesser response of the cortical vessels to hypoxia was suggested (33). However, anesthesia modifies many brain functions and blunts sympathetic nervous responses. We therefore recently studied the distribution of the brain blood flow response to hypoxia in unanesthetized cats. Hypoxia, induced by inhalation of 10% O_2, resulted in arterial P_{O_2} and P_{CO_2} values of 39 and 22 mmHg, respectively. Distribution of brain blood flow was assessed by the radiolabeled-microsphere technique, employing left ventricular injections.

The data are best appreciated by comparing cortical and brain stem flow (Fig. 2). The difference in euoxic flow represents different contents of gray matter in each region. A significant difference showed that the greatest increase in blood flow was in the medulla and the least response in the cortex. We believe that this is because in the unanesthetized state systemic hypoxia activates the sympathetic nervous system. The rostral-to-caudal distribution of sympathetic nerve terminals in the larger brain vessels is such that sym-

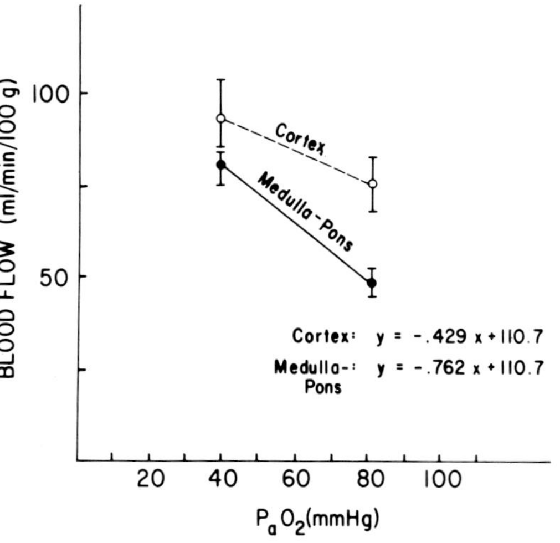

FIG. 2. Brain blood flow in cortex and medulla during euoxia and hypoxia in 6 unanesthetized cats. Responsiveness of medulla (*slope* of *line*) was significantly greater. Pa_{O_2}, partial pressure of O_2 in arterial blood.

pathetic activation would be expected to produce the greatest constriction and thereby the least increase in blood flow of the rostral (i.e., cortical) vessels. Thus the apparent greater functional vulnerability of the cortex, as compared to the brain stem, to acute hypoxia may be due at least partly to the lesser increase in cortical blood flow during hypoxia.

Implications of Brain Blood Flow Response to Acute Hypoxia for Ventilatory Control

The increase in brain stem blood flow during acute hypoxia may influence ventilation in two ways. First, the increase in flow increases O_2 delivery and preserves tissue PO_2. This appears to be accomplished quite effectively over a large range. For example, in unanesthetized goats given 1% CO in O_2 to breathe, whole brain blood flow increased so that O_2 delivery to the brain (product of O_2 content and blood flow) did not decline until carboxyhemoglobin reached levels of from 40% to 60% (8). Because brain stem flow increases more briskly than cortical flow in the unanesthetized animal and because cerebral flow is the major portion of whole brain flow in goats, brain stem O_2 delivery is probably kept at euoxic levels with even more severe hypoxia. This may be critically important in preserving ventilatory responses because, as St. John (29) has shown, brain stem respiratory neurons are affected by hypoxia heterogeneously, with some silenced by relatively mild hypoxia.

On the other hand, the respiratory control system probably compromises CO_2 homeostasis for the brisk brain stem vascular response to hypoxia. Because hypoxia in this range does not alter brain metabolic rate, the increase in blood flow should decrease tissue PCO_2 in the chemosensitive regions of the brain and thereby reduce the total afferent input into the respiratory controller. We have done preliminary studies to determine whether brain hypoxia could decrease ventilation in the unanesthetized goat. As in our previous studies, we

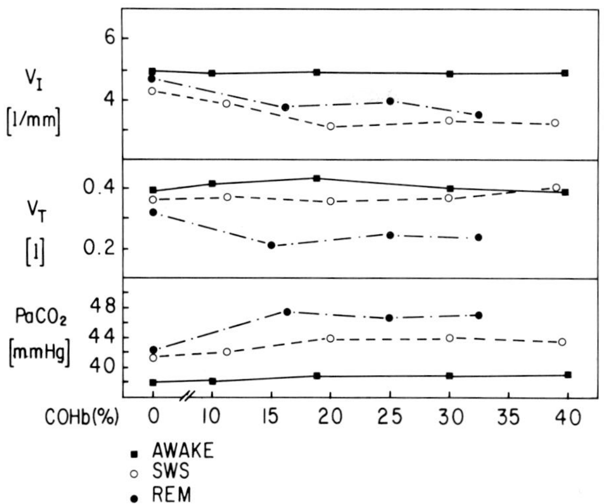

FIG. 3. Minute inspired ventilation (V_I), tidal volume (V_T), and Pa_{CO_2} during progressive carboxyhemoglobinemia in awake state and during slow-wave sleep (SWS) and rapid-eye-movement (REM) sleep for 1 goat. Progressive carboxyhemoglobinemia produced hypoventilation during sleep but not in awake state.

used CO to induce brain hypoxia without stimulation of the carotid bodies. Because the experiments were done in conjunction with our studies on ventilatory control during sleep, we administered 1% CO to the animals while awake and during both slow-wave sleep (SWS) and rapid-eye-movement (REM) sleep. Figure 3 shows that, although there was no effect in the awake state, carboxyhemoglobin levels of about 20% produced a reduction in ventilation of about 10% during sleep, especially during REM sleep. We believe that the lack of effect in the awake state reflects the predominance of the "wakefulness" stimulus to breathing in that state. However, during anesthesia or sleep, when the chemical stimuli of breathing dominate ventilatory control, the reduction of ventral medullary [H^+] is expressed as a decrease in ventilation.

It is difficult to determine from these studies whether washout of CO_2 from the central chemoreceptor by the increase in perfusion secondary to hypoxia is quantitatively significant in the control of breathing during an ascent to high altitude. However, the increase in brain blood flow during an ascent seems to abate with a time course similar to the phenomenon termed *ventilatory acclimatization* to altitude. The decline in brain blood flow and consequent increase of the tissue P_{CO_2} relative to that in arterial blood could be involved in the gradual increase in ventilatory stimulation, i.e., ventilatory acclimatization.

CHRONIC HYPOXIA

Whole brain blood flow returns close to sea-level values in sojourners within several days of ascent to altitude, that is, with a time course apparently parallel to that of ventilatory acclimatization (26). In that study the explanation offered (with supporting data) was that both ventilation and brain blood flow were controlled by the [H^+] of the brain interstitium and that the parallel changes reflected a return of the latter to a sea-level value from its initial alkaline value. Because more recent studies have questioned whether there is a return to sea-level $[H^+]$ of brain interstitial fluid coincident with acclimatization (6), this explanation must remain open to question. Whatever the mechanism, it is likely to reflect decreased brain tissue hypoxia, because reports of sojourners suggest that cortical function improves with acclimatization to altitude.

Natives of high altitude seem to have values for brain blood flow comparable to those at sea level. Figure 4 illustrates this and shows a sea-level value for brain blood flow at a sea-level value for blood hematocrit. In this study (28) brain blood flow was not measured but inferred from arteriovenous differences for O_2 content.

Two additional points are made in this figure. First, the cerebrovascular response to hypoxia appears to be intact. That is, when O_2 was administered to some of these subjects, the arteriovenous O_2 difference increased, implying a decrease in brain blood flow. Second, there was a strong inverse correlation between hematocrit and brain blood flow. This must at least in part reflect the dependence of brain blood flow on O_2 content of arterial blood; however, it probably also reflects the increase in blood viscosity with increasing hematocrit. The effect of hematocrit on blood viscosity and thereby on brain blood flow may be important to brain function in residents of high altitude. If all other factors (e.g., nutritional status) are equal, the most hypoxic subjects should have the highest hematocrits. Thus maladapted individuals could suffer brain hypoxia from both a lower-than-average arterial P_{O_2} and a suboptimal brain blood flow for that P_{O_2}. Such individuals, common in the Andes, have been described under the syndrome *chronic mountain sickness* (see chapt. 16). Occasional reports indicate that, in this setting, cognition is improved by

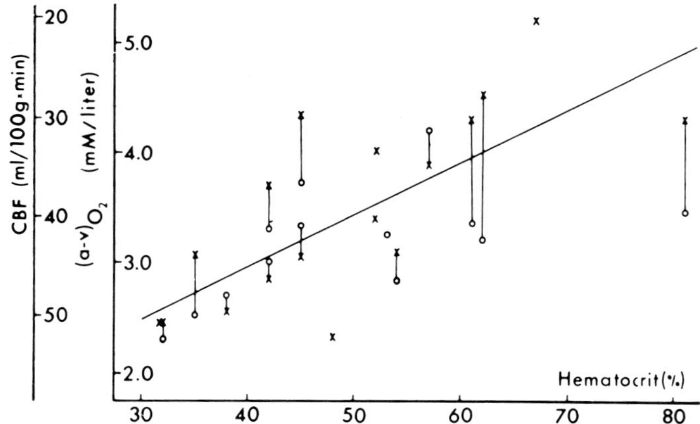

FIG. 4. Relationship between hematocrit and measured arteriovenous O_2 difference across the brain in high-altitude residents breathing either ambient air (○) or O_2 (×). Linear regression line through values obtained during O_2 breathing. *Left ordinate*, showing corresponding cerebral blood flow values, has been constructed by assuming a cerebral O_2 consumption of 3 ml·100 g^{-1}·min. [From Sørensen et al. (28).]

phlebotomy, suggesting that O_2 delivery to the brain may be improved by the decreased blood viscosity despite the concomitant lowering of O_2-carrying capacity of the blood.

BRAIN BLOOD FLOW DURING SLEEP

In the past 10 years the importance of studying the sleep state has become apparent. Clearly in normal individuals and in those with respiratory diseases, respiratory control and gas exchange are markedly modified during the various phases of sleep. Sojourners to high altitude have long appreciated this, frequently describing marked and disturbing periodic (Cheyne-Stokes) breathing during sleep at altitude. Because we believe that brain blood flow is an integral part of the overall respiratory control system, we have turned our attention to the regulation of brain blood flow during sleep.

Interest in brain blood flow during sleep extends as far back as 1881 when Mosso (21) attempted to test the hypothesis that sleep was due to brain hypoxia resulting from inadequate blood flow. Modern techniques have shown that during SWS brain blood flow increases to the extent expected from the concomitant increase of P_{CO_2} in arterial blood (20, 25). However, during REM sleep, brain blood flow increases rather briskly. The increase is more than would be expected from the associated elevation of arterial P_{CO_2} and the responsiveness of brain blood flow to CO_2 in awake or anesthetized animals. In the study by Reivich and co-workers (25) blood flow during REM sleep increased throughout the brain. These researchers believed the increase reflected an increase in metabolic rate of the brain; however, metabolic rate was not measured.

We have recently completed studies of brain blood flow during sleep with our model developed for study of unanesthetized goats. This model has allowed us to record flow continuously with an electromagnetic flow meter and to sample brain (primarily cortical) venous blood from the sagittal sinus. We have confirmed a large increase in brain blood flow during REM sleep, that is, greater than expected from the concomitant elevation of arterial P_{CO_2} based on data from animals in SWS or awake animals inhaling CO_2-enriched gas. This is illustrated in Figure 5, which depicts on a continuous time base the transition from SWS to REM sleep. Virtually without time lag, brain blood flow increased as REM sleep started. This abrupt onset of the flow response suggested that it might not be due to increased tissue metabolism. We tested this directly by measuring arteriovenous differences for O_2 across the brain. The arteriovenous difference narrowed (Fig. 6), and the calculated O_2 consumption was unchanged from that of the awake state, indicating that, contrary to the supposition by Reivich and co-workers (25), brain hypermetabolism does not cause the increased brain blood flow during REM sleep.

Additional studies revealed that not only was euoxic brain blood flow enhanced in REM sleep but the blood flow response to hypoxia was also enhanced. We used progressive carboxyhemoglobinemia to induce brain hy-

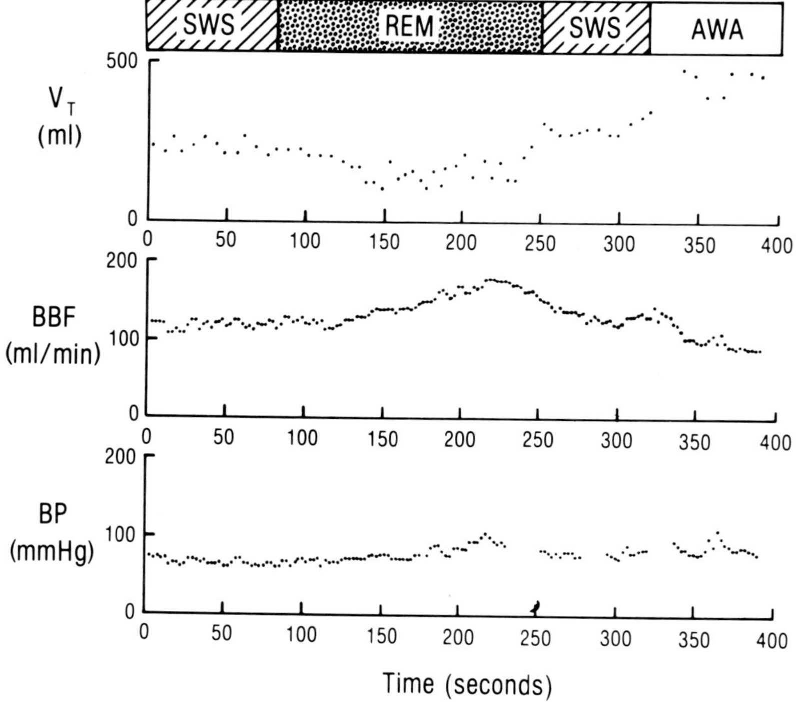

FIG. 5. Systemic blood pressure (BP), brain blood flow (BBF), and V_T during SWS, REM sleep, and awake (AWA) state. Decrease in V_T was coincident with increase in BBF. [From Santiago et al. (25a), by copyright permission of the American Society for Clinical Investigation.]

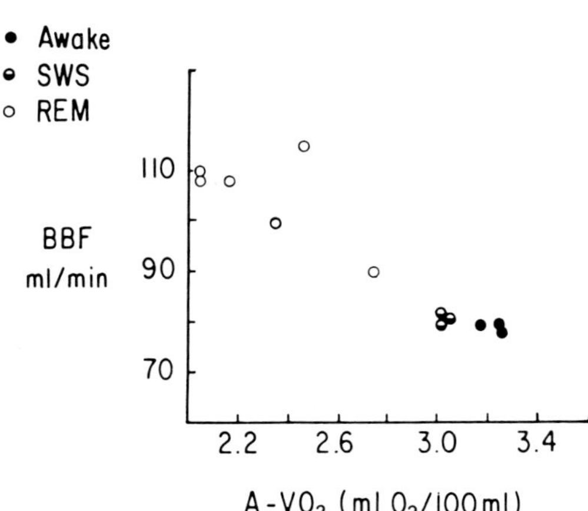

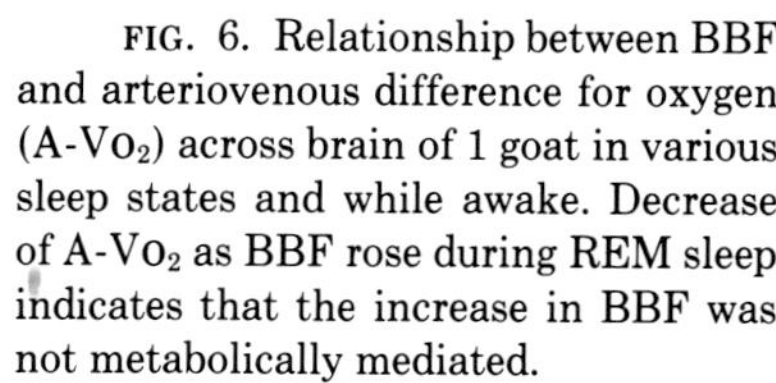

FIG. 6. Relationship between BBF and arteriovenous difference for oxygen ($A\text{-}V_{O_2}$) across brain of 1 goat in various sleep states and while awake. Decrease of $A\text{-}V_{O_2}$ as BBF rose during REM sleep indicates that the increase in BBF was not metabolically mediated.

poxia without stimulation of peripheral arterial chemoreceptors. Figure 7 shows the increase in brain blood flow during inhalation of 1% CO in 40% O_2. The response during SWS was similar to the response during wakefulness but was offset to a slightly higher level because of the elevation of arterial P_{CO_2} during sleep. In contrast brain blood flow was markedly enhanced in response to hypoxia during REM sleep.

The mechanism of enhanced brain blood flow and brain blood flow

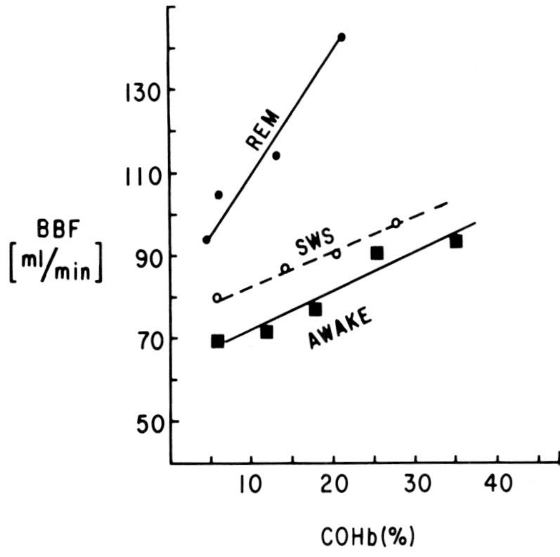

FIG. 7. Relationship between BBF and carboxyhemoglobin level during inhalation of 1% CO in 40% O_2 during various sleep states and while awake. The BBF response to hypoxia was much greater during REM sleep.

responsiveness to hypoxia during REM sleep has not yet been elucidated. However, if (as we believe) it is not related to an increased metabolic rate, the implications for ventilatory control are substantial. The inescapable conclusion is that tissue P_{CO_2} is decreased, that is, moved closer to the level in arterial blood during REM sleep. If this occurs in the region of the central chemoreceptors, it should decrease chemoreceptor stimulation. During REM sleep in a hypoxic environment, P_{CO_2} at the central chemoreceptor would be reduced even further, probably quite close to its threshold value. As shown in chapter 10, reduction of the CO_2 stimulus to and below its threshold level at the central chemoreceptor is probably necessary for the generation of periodic (Cheyne-Stokes) breathing during sleep at altitude.

SUMMARY

A brisk increase in brain blood flow with acute hypoxia is an important component of the protective physiological responses that occur on ascent to high altitude. Apparently the response is regionally uneven in unanesthetized animals, the greatest increase in flow occurring in the brain stem. This may protect basic control systems mediated by the central nervous system. The less brisk increase of cortical flow in acute hypoxia must be partly responsible for the greater functional vulnerability of the cortex to hypoxia.

We propose that the relatively long time constant of the brain blood flow response to acute hypoxia and the postulated CO_2 washout from chemosensitive areas of the brain stem modulate ventilation during ascent to altitude. However, there is no direct proof of this. Hypoxia causes an unexpectedly brisk increase of brain blood flow during REM sleep, which seems unrelated to changes in brain metabolism. Such an arrangement has the potential for marked modification of P_{CO_2} at the chemosensitive site during REM sleep at altitude.

When a stay at altitude is extended, the brain blood flow response to acute hypoxia abates quickly with a time course similar to that of ventilatory acclimatization. It is not clear whether this implies a causal relationship or reflects control by the same or temporally concordant variables.

The expert editorial assistance of Marcella Spioch is appreciated.

Some of the work reported in this paper was supported by National Institutes of Health Grants HL-16022, HL-23315, and HL-07467 and a grant from the New Jersey Thoracic Society.

REFERENCES

1. ASTRUP, J., D. HEUSER, N. A. LASSEN, K. NILSSON, K. NORBERG, AND B. K. SIESJÖ. Evidence against H^+ and K^+ as the main factors in the regulation of cerebral blood flow during epileptic discharges, acute hypoxemia, amphetamine intoxication and hypoglycaemia. A microelectrode study. In: *Ionic Actions on Vascular Smooth Muscle*, edited by E. Betz. Berlin: Springer-Verlag, 1976, p 110–115.
2. CAMERON, I. B. The chemical control of the cerebral circulation. *Clin. Sci. Mol. Med.* 52: 549–554, 1977.
3. CHAPMAN, R. W., T. V. SANTIAGO, AND N. H. EDELMAN. Effects of graded reduction of brain blood flow on ventilation in unanesthetized goats. *J. Appl. Physiol.: Respirat. Environ. Exercise Physiol.* 47: 104–111, 1979.
4. CHAPMAN, R. W., T. V. SANTIAGO, AND N. H. EDELMAN. Effects of graded reduction of brain blood flow on chemical control of breathing. *J. Appl. Physiol.: Respirat. Environ. Exercise Physiol.* 47: 1289–1294, 1979.
5. CHAPMAN, R. W., T. V. SANTIAGO, AND N. H. EDELMAN. Brain hypoxia and control of breathing: neurochemical control. *J. Appl. Physiol.: Respirat. Environ. Exercise Physiol.* 49: 497–505, 1980.
6. DEMPSEY, J. A., AND H. V. FORSTER. Mediation of ventilatory adaptations. *Physiol. Rev.* 62: 262–346, 1982.
7. DOBLAR, D. D., B. G. MIN, R. W. CHAPMAN, E. R. HARBACK, W. WELKOWITZ, AND N. H. EDELMAN. Dynamic characteristics of cerebral blood flow response to sinusoidal hypoxia. *J. Appl. Physiol.: Respirat. Environ. Exercise Physiol.* 46: 721–729, 1979.
8. DOBLAR, D. D., T. V. SANTIAGO, AND N. H. EDELMAN. Correlation between ventilatory and cerebrovascular responses to inhalation of CO. *J. Appl. Physiol.: Respirat. Environ. Exercise Physiol.* 43: 455–462, 1977.
9. EDVINSSON, L. Neurogenic mechanisms in the cerebrovascular bed: autonomic nerves, amine receptors, and their effects on cerebral blood flow. *Acta Physiol. Scand. Suppl.* 427: 1–35, 1975.
10. EDVINSSON, L., AND E. T. McKENZIE. Amine mechanisms in the cerebral circulation. *Pharmacol. Rev.* 28: 275–348, 1977.
11. HARPER, A. M., V. D. DESHMUKH, J. O. ROWAN, AND W. B. JENNETT. The influence of sympathetic nervous activity on cerebral blood flow. *Arch. Neurol.* 27: 1–6, 1972.
12. HEISTAD, D. D., M. L. MARCUS, J. C. EHRHARDT, AND F. M. ABBOUD. Effect of stimulation of carotid chemoreceptors on total and regional cerebral flow. *Circ. Res.* 38: 20–25, 1976.
13. JAMES, I. M., R. A. MILLAR, AND M. J. PURVES. Observations on the extrinsic neural control of cerebral blood flow in the baboon. *Circ. Res.* 25: 77–93, 1969.
14. JONES, M. D., JR., R. J. TRAYSTMAN, M. A. SIMMONS, AND R. A. MOLTENI. Effects of changes in arterial O_2 content on cerebral blood flow in the lamb. *Am. J. Physiol.* 240 (*Heart Circ. Physiol.* 9): H209–H215, 1981.
15. KONTOS, H. A. Regulation of the cerebral circulation. *Annu. Rev. Physiol.* 43: 397–407, 1981.
16. KONTOS, H. A., E. P. WEI, R. M. NAVARI, J. E. LEVASSEUR, W. I. ROSENBLUM, AND J. L. PATTERSON, JR. Response of cerebral arteries and arterioles to acute hypotension and hypertension. *Am. J. Physiol.* 234 (*Heart Circ. Physiol.* 3): H371–H383, 1978.
17. KUSCHINSKY, W., AND M. WAHL. Local chemical and neurogenic regulation of cerebral vascular resistance. *Physiol. Rev.* 58: 656–689, 1978.
18. LASSEN, N. A. Brain extracellular pH: the main factor controlling cerebral blood flow. *Scand. J. Clin. Lab. Invest.* 22: 247–251, 1968.
19. McMILLAN, V., L. G. SALFORD, AND B. K. SIESJÖ. Metabolic state and blood flow in rat cerebral cortex, cerebellum and brainstem in hypoxic hypoxia. *Acta Physiol. Scand.* 92: 103–113, 1974.
20. MANGOLD, R., L. SOKOLOFF, E. CONNOR, J. KLEINERMAN, P. G. THERMAN, AND S. S. KETY. The effects of sleep and lack of sleep on the cerebral circulation and metabolism of normal young men. *J. Clin. Invest.* 34: 1092–1100, 1955.
21. MOSSO, A. Über den kreislauf des blutes in menschlichen gehirn. *Leipz. Viet.* 1881.
22. NILSSON, B., K. NORBERG, C. H. NORDSTROM, AND B. K. SIESJÖ. Influence of hypoxia and hypercapnia on CBF in rats. In: *Blood Flow and Metabolism in the Brain*, edited by M. Harper, B. Jennett, D. Miller, and J. Rowan. Edinburgh: Churchill Livingstone, 1975, p. 9.19–9.23.
23. NILSSON, B., S. REHNCRONA, AND B. K. SIESJÖ. Coupling of cerebral metabolism and blood flow in epileptic seizures, hypoxia and hypoglycemia. In: *Cerebral Vascular Smooth Muscle and Its Control*, edited by M. Purves. Amsterdam: Excerpta Med., 1978, p. 199–214. (Ciba Found. Symp. 56.)
24. PONTE, J., AND M. J. PURVES. The role of the carotid body chemoreceptors and carotid sinus baroreceptors in the control of cerebral blood vessels. *J. Physiol. London* 237: 315–340, 1974.
25. REIVICH, M., G. ISAACS, E. EVARTS, AND S. S. KETY. The effect of slow wave sleep and REM sleep on regional cerebral blood flow in cats. *J. Neurochem.* 15: 301–306, 1968.

25a. SANTIAGO, T. V., E. GUERRA, J. A. NEUBAUER, AND N. H. EDELMAN. Correlation between ventilation and brain blood flow during sleep. *J. Clin. Invest.* 73: 497–506, 1984.

26. SEVERINGHAUS, J. W., H. CHIODI, E. I. EGER II, B. BRANDSTATER, AND T. F. HORNBEIN. Cerebral blood flow in man at high altitude. *Circ. Res.* 19: 274–282, 1966.
27. SIESJÖ, B. K., L. BERNTMAN, AND B. NILSSON. Regulation of microcirculation in the brain. *Microvasc. Res.* 19: 158–170, 1980.
28. SØRENSEN, S. C., N. A. LASSEN, J. W. SEVERING-

HAUS, J. COUDERT, AND M. P. ZAMORA. Cerebral glucose metabolism and cerebral blood flow in high-altitude residents. *J. Appl. Physiol.* 37: 305–310, 1974.

29. ST. JOHN, W. M. AND S. C. WANG. Response of medullary respiratory neurons to hypercapnia and isocapnic hypoxia. *J. Appl. Physiol.: Respirat. Environ. Exercise Physiol.* 43: 812–821, 1977.
30. TRAYSTMAN, R. J., R. S. FITZGERALD, AND S. LOSCUTOFF. Cerebral circulatory responses to arterial hypoxia in normal and chemodenervated dogs. *Circ. Res.* 42: 649–657, 1978.
31. WADE, J. P., G. H. DU BOULAY, J. MARSHALL, T. C. PEARSON, R. W. RUSSELL, J. A. SHIRLEY, L. SYMON, G. WETHERLEY-MEIN, AND E. ZILKHA. Cerebral blood flow, haematocrit and viscosity in subjects with a high oxygen affinity haemoglobin variant. *Acta Neurol. Scand.* 61: 210–215, 1980.
32. WEI, E. P., AND H. A. KONTOS. Responses of cerebral arterioles to increased venous pressure. *Am. J. Physiol.* 243 (*Heart Circ. Physiol.* 12): H442–H447, 1982.
33. WEISS, H., AND N. H. EDELMAN. Effect of hypoxia on small vessel blood content of rabbit brain. *Microvasc. Res.* 12: 305–315, 1976.
34. WINN, H. R., R. RUBIO, AND R. M. BERNE. Brain adenosine concentration during hypoxia in rats. *Am. J. Physiol.* 241 (*Heart Circ. Physiol.* 10): H235–H242, 1981.

10

Hypoxic Versus Hypocapnic Effects on Periodic Breathing During Sleep

ANNE BERSSENBRUGGE, JEROME DEMPSEY,
AND JAMES SKATRUD

John Rankin Laboratory of Pulmonary Medicine, Department of Preventive Medicine, University of Wisconsin; and William S. Middleton Memorial Veterans Administration Hospital, Madison, Wisconsin

PERIODIC BREATHING DURING SLEEP in healthy persons at high altitudes was first documented in the 1890s (16). Although additional studies in humans (1a, 22, 29–31) have demonstrated periodic breathing at high altitudes, the mechanism of hypoxia-induced periodic breathing and its relationship to sleep remain unclear. In 1909 Douglas and Haldane (7) induced periodic breathing in normal humans after periods of acute voluntary hyperventilation and suggested the importance of both hypoxia and hypocapnia on its occurrence. Recent studies based on computer modeling and anesthetized-animal experiments propose that periodic breathing reflects instability in the chemical feedback control of ventilation (3–6, 13, 17, 18). In the series of studies described in this chapter, we combined sleep and hypoxia in normal humans to test for possible mechanisms involved in the genesis of periodic breathing (for more detailed information, see ref. 1). Under these conditions the ventilatory control system may be altered by the following: *1*) neurophysiological changes in the higher central nervous system associated with individual changes in sleep state; *2*) hypoxia itself, which enhances peripheral feedback, and the attendant hyperventilation and hypocapnic alkalosis, which decrease central chemoreceptor feedback; and *3*) a direct effect of hypoxia on the brain (i.e., cerebral hypoxic depression and cerebral alkalosis) (11, 25) or a direct effect of hypoxia or hypocapnia on upper airway resistance (8). Our purpose was to both quantitatively describe the nature of hypoxia-induced periodic breathing on a breath-to-breath basis and investigate how the sleep state, hypoxia per se, and hypocapnic alkalosis secondary to hypoxia may contribute to hypoxia-induced ventilatory instability and periodic breathing.

METHODS

Six healthy male subjects (ages 22–34 yr) were studied during wakefulness and sleep both in normoxia [mean barometric pressure (P_B) = 740 mmHg] and in hypobaric hypoxia in an altitude chamber (P_B = 455 mmHg; at 0.5- to 9.0-h duration of hypoxia exposure). Subjects were studied predominantly during air breathing. Intermittently, O_2 and varied mixtures of CO_2-enriched air were administered via nasal cannulas for intervals of 10–15 min. During administration of CO_2 in hypoxia, additional N_2 was bled into the inspired gas mixture to maintain constant mean arterial O_2 saturation (Sa_{O_2}).

Techniques for monitoring sleep stage, Sa_{O_2}, arterial pH and partial pressure of CO_2 in arterial blood (Pa_{CO_2}), and ventilation (inductance plethysmography) have been described (10, 14, 21, 27). All ventilatory measurements were quantitated according to sleep stage; only sections in which sleep stage was constant for more than 3 min were included for analyses to avoid the effects of fluctuating sleep state that previously have been associated with periodic breathing in normal man (2). Ventilation was analyzed for mean values of expired minute ventilation [$\dot{V}_E$ = sum of tidal volumes (V_T)/min] and frequency (f = sum of breaths/min) as well as for breath-to-breath measurements of $\dot{V}_E$, f, V_T, total breath cycle time (T_T, measured from onset of inspiration to onset of subsequent inspiration), inspiratory duration (T_I), expiratory duration (T_E), mean inspiratory flow (V_T/T_I), and "duty cycle" (T_I/T_T). Expiratory duration was further subdivided into that portion during which expiratory flow occurred ($T_{E_{short}}$) and that portion during which no flow occurred, i.e., the expiratory pause (EP). The stability of the breathing pattern was assessed by the variation about the mean levels of the individual volume and timing components of the breath (expressed graphically by frequency-distribution histograms) and by the number and length of EPs of ≥5-s duration, which we defined as apnea. Ventilatory measurements from stage II, III, and IV sleep were subsequently combined to represent non-rapid-eye-movement (non-REM) sleep, because no differences were found among them.

RESULTS AND DISCUSSION

Effect of Sleep State on Ventilation in Normoxia

The effects of sleep on breathing in normoxia are summarized for all subjects in Figure 1. Compared to wakefulness, during non-REM sleep $\dot{V}_E$ decreased 12% and was accompanied by a mild respiratory acidosis (Pa_{CO_2} increased 2.2–5.2 mmHg), whereas mean Sa_{O_2} was unchanged (range 96–98%). This relative hypoventilation during non-REM sleep, previously reported in normal subjects (2, 12, 27), occurred in the absence of any significant change in V_T, V_T/T_I, or breathing cycle timing. Further significant changes in the magnitude of any of these variables were not observed during REM sleep.

Examples of breathing patterns for non-REM and REM sleep during

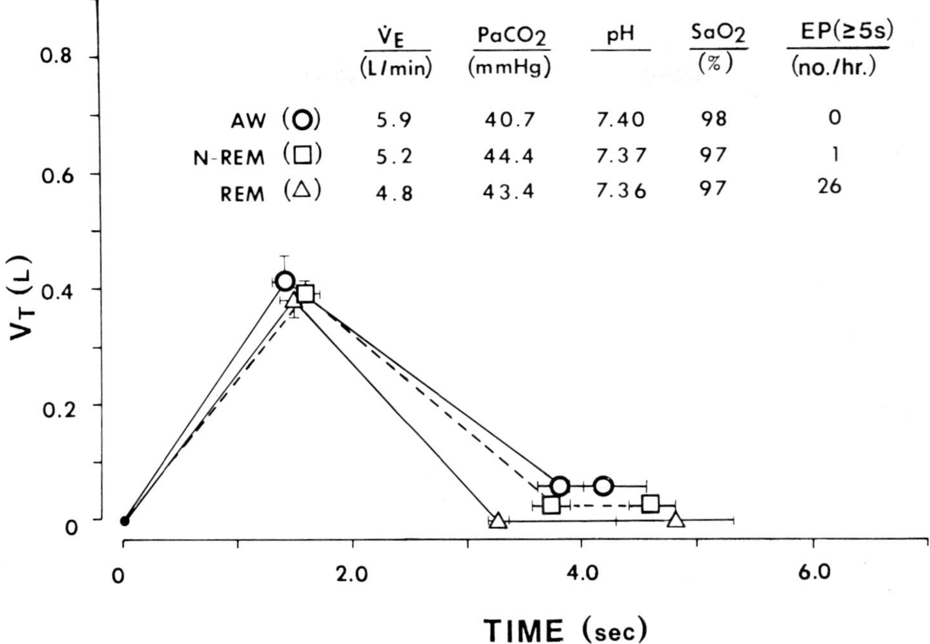

FIG. 1. Average spirograms and mean ventilatory measurements from all subjects during wakefulness (AW), rapid-eye-movement (REM) sleep, and non-REM (N-REM) sleep in normoxia. T_E has been subdivided into 2 intervals: when expiratory flow is present ($T_{E_{short}}$) and expiratory pause (EP = $T_E - T_{E_{short}}$). Means ± SE.

normoxia are illustrated for one subject in Figure 2 (*upper panel*), and the effect of sleep state on breath-to-breath variability is summarized for all subjects by the frequency distributions shown in Figure 3. During non-REM sleep the breathing pattern was generally consistent from breath to breath with respect to both volume and timing, and apneas (EP ≥ 5 s) were rarely observed (≤1/h). In contrast, during REM sleep the breathing pattern was frequently characterized by random erratic variations in both the volume and timing of breaths, as described in a variety of species (2, 24, 28). Thus the distributions of V_T and calculated breath-to-breath $\dot{V}_E$ (Fig. 3) show slightly wider ranges during normoxic REM sleep than during normoxic non-REM sleep.

Effect of Sleep State on Ventilation in Hypoxia

The effects of sleep on breathing in hypoxia are summarized for all subjects in Figure 4. During hypoxia mean Sa_{O_2} was decreased and hyperventilation and respiratory alkalosis ($\Delta Pa_{CO_2} = -7$ mmHg) occurred during wakefulness and all sleep stages relative to their normoxic values (cf. Fig. 4 and Fig. 1).

The most striking effect of hypoxia was periodic breathing, which was the predominant breathing pattern for all stages of non-REM sleep in all subjects

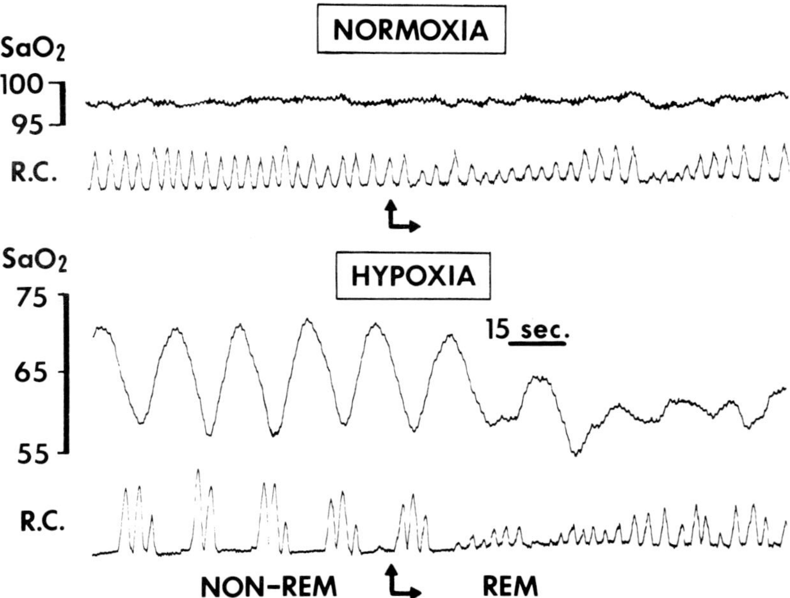

FIG. 2. Tracings of Sa_{O_2} and rib cage (RC) movements from 1 subject. Spontaneous transitions from non-REM to REM sleep (at *arrows*) in normoxia (*upper panel*) and in hypoxia (*lower panel*).

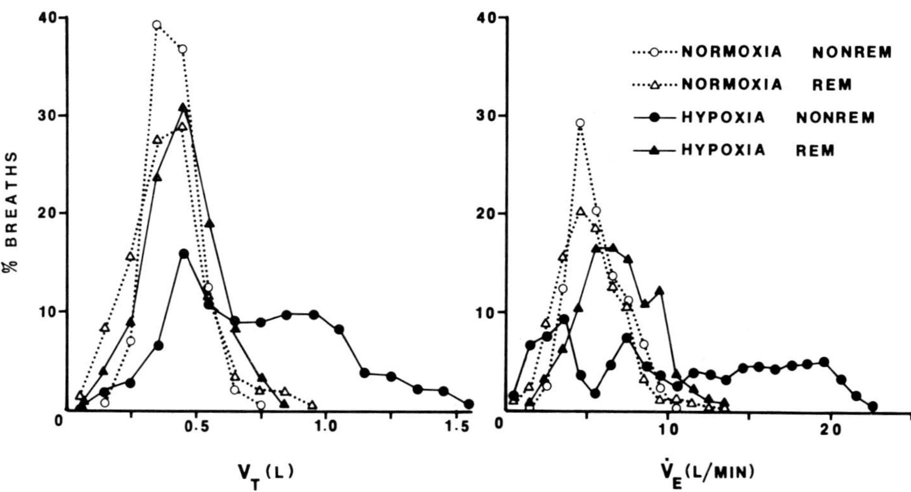

FIG. 3. Frequency distributions of VT and calculated breath-to-breath minute ventilation ($\dot{V}$E) during sleep in normoxia and hypoxia, averaged from all subjects.

but was rarely observed during wakefulness or REM sleep (see, e.g., Fig. 2, *lower panel*). The periodic breathing of non-REM sleep showed the following characteristics: *1*) repetitive breathing oscillations of reproducible cycle length (21.2 ± 1.8 s; mean ± SE), which were associated with large swings in Sa_{O_2} ($\bar{x}$ = 66%; $\bar{x}_{min}$ = 54%; $\bar{x}_{max}$ = 76%); *2*) clusters of 2–5 breaths of highly variable

V_T and breath-to-breath $\dot{V}_E$, which alternated regularly with a prolonged EP of the last breath of each cluster (apnea); and *3*) nonobstructive or "central" origins for these apneas. The proportion of time spent in apnea accounted for almost half of the cycle length of the periodic-breathing episode. The mean number of apneas per hour was 145 ± 15 (SE) (Fig. 4), with an average duration of 10.8 s (range 5–18 s).

Breathing "clusters" during periodic breathing had quite different timing and pattern characteristics from those observed during the rhythmic (nonperiodic) breathing of wakefulness or REM sleep in hypoxia (Fig. 4): *1*) V_T and V_T/T_I were increased (indicating augmented inspiratory effort during breathing clusters); *2*) T_T was increased (due to prolonged EP of last breath of cluster); *3*) T_I and $T_{E_{short}}$ were unchanged.

In hypoxia the breath-to-breath variability was greater during non-REM than REM sleep, reflecting the periodic pattern of breathing. During REM sleep, however, the random variations in volume and timing of breaths typical of the breathing pattern during normoxic REM sleep were unaffected by hypoxic exposure (Fig. 2). Thus the distributions of V_T and breath-to-breath $\dot{V}_E$ (Fig. 3) show the widest ranges during hypoxic non-REM sleep, whereas during hypoxic REM sleep the ranges of distribution are similar to those observed during normoxic REM sleep (although mean V_T and $\dot{V}_E$ are greater in hypoxia). The absence of periodic breathing during REM sleep in normal subjects at altitude has also been reported by others (22, 31).

These studies demonstrate that the state-dependent effect of hypoxia on

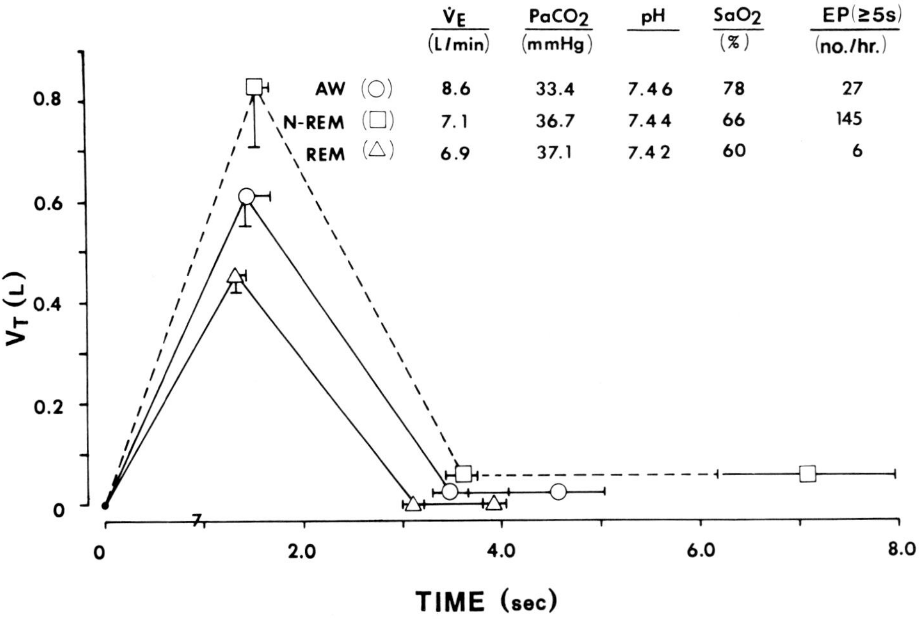

FIG. 4. Average spirograms and mean ventilatory measurements from all subjects during wakefulness, non-REM sleep, and REM sleep in hypoxia. Awake measurements were collected during 7–9 h of hypoxic exposure after the sleep portion of the study.

periodic breathing was confined to non-REM sleep, although equivalent or greater levels of hypoxia or hypoxia-induced respiratory alkalosis occurred during wakefulness or REM sleep. These results support the proposal that, compared to non-REM sleep, dependence on classic respiratory afferent stimuli decreases in wakefulness and REM sleep and that respiratory rhythm is maintained, when awake, by the "wakefulness drive" to breathe and, during REM sleep, by the intrinsic nonrespiratory neural events of the higher central nervous system (2, 9, 20, 23). Thus during wakefulness or REM sleep these nonchemical influences on inspiratory activity apparently provide sufficient stimuli to maintain respiratory cycle rhythm despite the alterations in chemical afferent input that occur in hypoxia.

Contributions of Hypoxia and Hypocapnic Alkalosis to Genesis of Periodic Breathing During Non-REM Sleep

To assess the relative importance of hypoxia per se and hypocapnic alkalosis secondary to hypoxia on the genesis of periodic breathing, we manipulated both of these factors, alone and in combination, and measured their effects on breathing pattern.

The effect of O_2 administration in hypoxia is illustrated for one subject in Figure 5, and mean steady-state data from all subjects are summarized in Figures 6 and 7. Augmentation of the fractional concentration of O_2 in dry inspired gas ($F_{I_{O_2}}$) during periodic breathing gradually increased Sa_{O_2} (time to 90% Sa_{O_2} = 1.2–4.0 min) and eliminated periodic breathing (Fig. 5). This transition to a regular pattern of breathing with increasing Sa_{O_2} was charac-

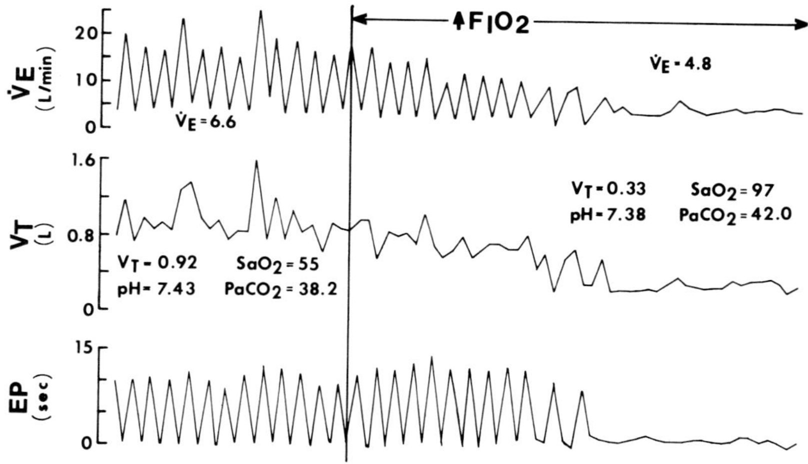

FIG. 5. Breath-to-breath measurements of calculated V̇E, VT, and duration of EP for 1 trial of O_2 administration during non-REM sleep in hypoxia. $F_{I_{O_2}}$ was increased at *vertical line.* During breath-to-breath measurements (74 breaths; 9.0-min period) non-REM sleep stages II and III were maintained. Periodic breathing is reflected in breath-to-breath oscillations in V̇E and VT and prolonged EP values, which are substantially reduced after O_2 administration. All measurements listed are mean steady-state values with same units as defined in Fig. 1. Also note that mean breath-to-breath V̇E overestimates actual V̇E, which was measured as sum of VT/min.

terized by *1*) a progressive decrease in VT and breath-to-breath V̇E, *2*) an initial lengthening of apneas followed by shortening and disappearance when $Sa_{O_2} > 91\%$, and *3*) elimination of cyclical oscillations in breath-to-breath V̇E and VT within 0.3–2.0 min of the cessation of apneas. The stabilizing effects of O_2 administration on breath-to-breath variability during hypoxia are summarized for all trials in Figure 6, which shows a substantial reduction in the ranges of VT and calculated breath-to-breath V̇E distributions during augmented FI_{O_2}. Steady-state values obtained 5 min after the onset of increased FI_{O_2} show that mean Sa_{O_2} increased from 63% to 97%, V̇E decreased (−1.3 ± 0.2 liters/min), Pa_{CO_2} increased (+4.2 ± 0.4 mmHg), and the breathing pattern was altered by changes in both inspiratory effort (decreased VT/TI and VT) and respiratory cycle timing (decreased EP and increased TI/TT) (Fig. 7).

Thus during the stabilization of breathing pattern with O_2 administration there was a concomitant increase in Pa_{CO_2}, suggesting that this stabilization may not simply be caused by the reversal of hypoxia per se but may also be caused by the reversal of hypocapnia. To test this possibility we administered exogenous CO_2 during periodic breathing in hypoxia while Sa_{O_2} was held constant. An example is illustrated for one subject in Figure 8, and mean steady-state data are summarized in Figures 9 and 10. Augmented FI_{CO_2} increased Pa_{CO_2} from 0.3 to 2.8 mmHg, decreased arterial pH, and increased V̇E, whereas mean Sa_{O_2} was unchanged (Fig. 9). Augmented FI_{CO_2} eliminated periodic breathing and stabilized the breathing pattern rapidly and reversibly (Fig. 8). Apneas were eliminated within 15 s of the onset of CO_2 administration, and cyclic oscillations in VT and breath-to-breath V̇E were eliminated within 1–2 min. Conversely apneas and periodic breathing returned within 30 s of the termination of CO_2 administration. The stabilizing effects of CO_2 administration on breath-to-breath variability during hypoxia are summarized for all trials in Figure 10, which shows a significant narrowing of the ranges of VT

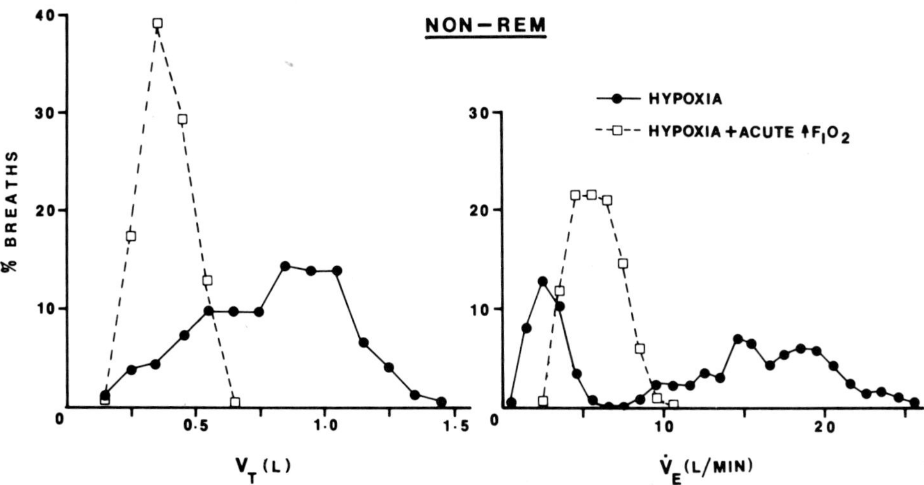

FIG. 6. Frequency distribution of VT and calculated breath-to-breath V̇E during hypoxia and during O_2 administration in hypoxia (averaged from 8 trials, 5 subjects).

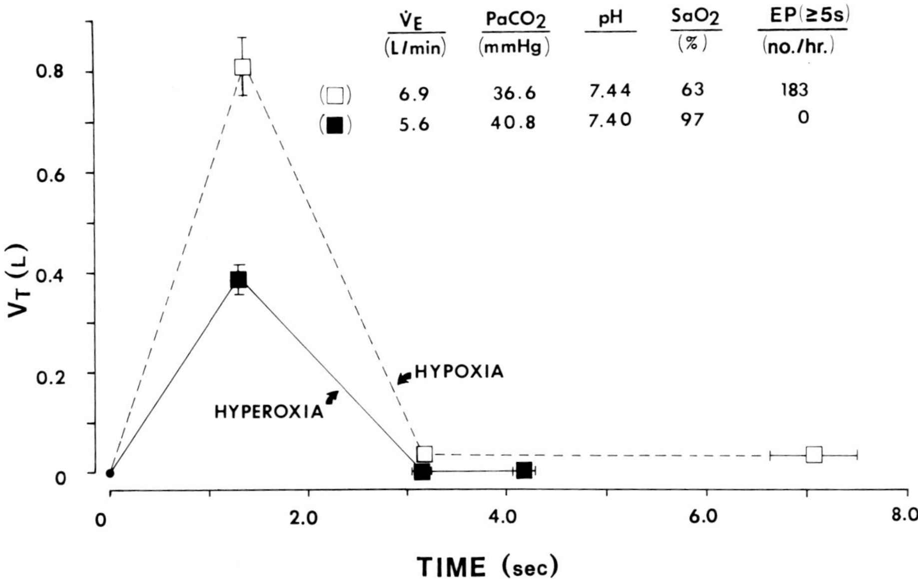

FIG. 7. Average spirograms and mean ventilatory measurements in non-REM sleep during hypoxia (□) and during O_2 administration in hypoxia (■). Values are means ± SE from 8 trials of increased F_{IO_2} in 5 subjects. Hypoxia measurements were collected within ±1.5 h of each O_2 administration.

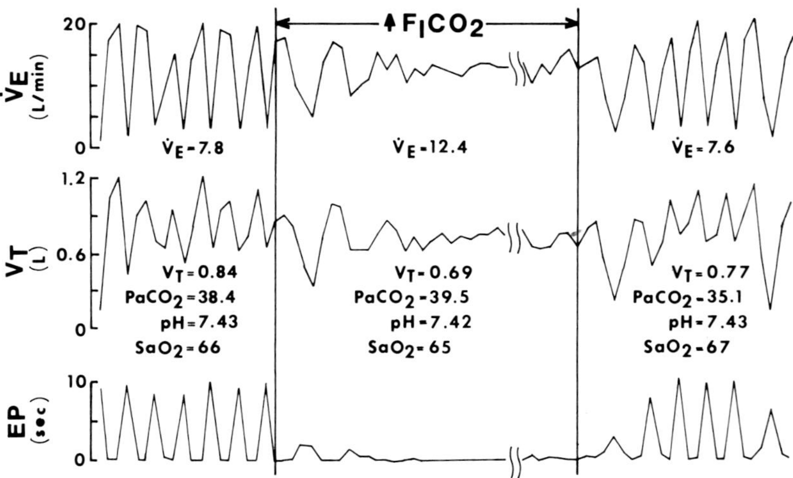

FIG. 8. Breath-to-breath measurements of calculated $\dot{V}_E$, V_T, and duration of EP for 1 trial of CO_2 administration during non-REM sleep in hypoxia. CO_2 added at *1st vertical line* and removed at *2nd vertical line.* During breath-to-breath measurements (75 breaths; 6.0-min period) non-REM sleep stage II was maintained. Periodic breathing is reflected in breath-to-breath oscillations in $\dot{V}_E$ and V_T and prolonged EP values, which are substantially reduced during CO_2 administration. Measurements listed are steady-state values with same units as defined in Fig. 1.

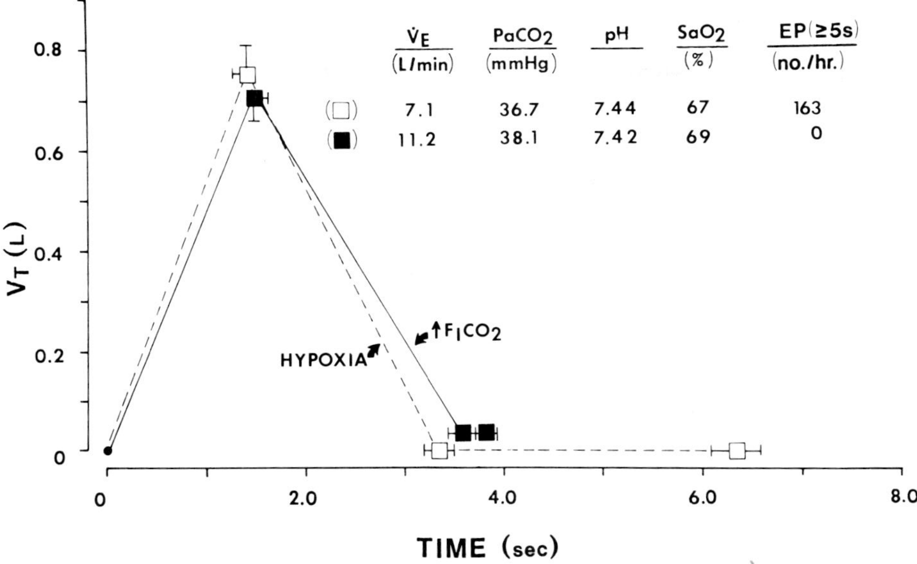

FIG. 9. Average spirograms and mean ventilatory measurements in non-REM sleep during hypoxia (□) and during CO_2 administration in hypoxia (■). Values are means ± SE from 9 trials of augmented FI_{CO_2} in 4 subjects. Hypoxia measurements were collected within ±30 min of each CO_2 administration.

and calculated breath-to-breath V̇E distributions during augmented FI_{CO_2}. The stabilizing effect of exogenous CO_2 on the breathing pattern occurred with a <2-mmHg increase in Pa_{CO_2}. Thus hypocapnic alkalosis persisted compared to values in normoxic non-REM sleep (cf. Fig. 9 and Fig. 1). The major effect of augmented CO_2 on breathing pattern in hypoxia was on respiratory cycle timing and specifically the elimination of apneas (decreased EP and increased TI/TT), whereas VT and VT/TI were unchanged (Fig. 9). These results contrast with our findings in normoxia in which augmented CO_2 increased V̇E mainly by increasing VT and VT/TI, with no effect on respiratory cycle timing. Analogous results have been reported in preterm infants during sleep by Kalapesi et al. (15). They found that augmented CO_2 stabilized breathing pattern when it was administered during periodic breathing mainly by changing respiratory cycle timing (i.e., decreased TE and increased f), whereas administering CO_2 during rhythmic breathing altered the pattern mainly by increasing VT, with little effect on respiratory cycle timing.

Further evidence supporting a role of hypocapnic alkalosis in hypoxia-induced periodic breathing was obtained from studies in which acute hypoxia was induced during non-REM sleep by lowering FI_{O_2} (12.5%) with a ventilation hood canopy while monitoring time course changes in end-tidal PO_2 (PET_{O_2}) and PCO_2 (PET_{CO_2}) and ventilatory pattern. Acute induction of hypocapnic hypoxia caused periodic breathing and apneas within 8 min of the onset of hypoxia, which correlated with both a decrease in PET_{O_2} and a progressive decrease in PET_{CO_2} (−4 mmHg). However, with acute induction of isocapnic

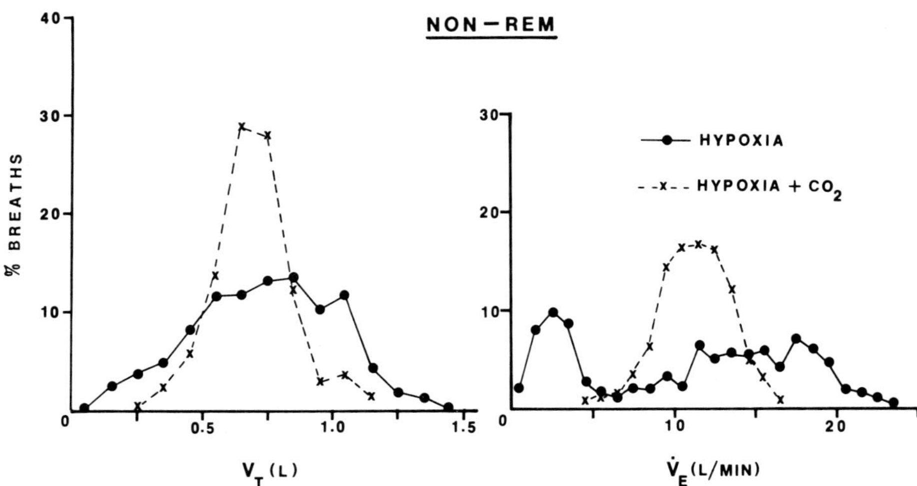

FIG. 10. Frequency distributions of VT and calculated breath-to-breath $\dot{V}$E during hypoxia and during CO_2 administration in hypoxia (averaged from 9 trials in 4 subjects).

hypoxia (during which $P_{ET_{CO_2}}$ was held at normocapnic levels by augmenting $F_{I_{CO_2}}$) the breathing pattern remained rhythmic for the 20-min duration of isocapnic hypoxia. Only after $P_{ET_{CO_2}}$ was allowed to fall (4-mmHg decrease) in the absence of any change in Sa_{O_2} did periodic breathing develop.

Clearly hypocapnic alkalosis is necessary in the production of hypoxia-induced periodic breathing during non-REM sleep. To investigate possible mechanisms, we used passive positive-pressure hyperventilation to induce hypocapnia in normal subjects in backgrounds of both hyperoxia and hypoxia (26). During wakefulness we were unable to elicit consistent apneas in either hyperoxia or hypoxia even with passive reductions in $P_{ET_{CO_2}}$ as large as 15 mmHg. In contrast, during non-REM sleep in either background condition, apneas were easily and consistently produced with considerably smaller reductions in $P_{ET_{CO_2}}$. During non-REM sleep in hyperoxia, apneas occurred when $P_{ET_{CO_2}}$ was passively reduced 3–6 mmHg below the eupneic levels of $P_{ET_{CO_2}}$ normally obtained in non-REM sleep, i.e., at $P_{ET_{CO_2}}$ values within 1–2 mmHg of their normal waking level. Furthermore the duration of apnea was significantly correlated with the degree of hypocapnia. After the apnea the recovery of spontaneous breathing showed a gradual return of VT toward normal values as $P_{ET_{CO_2}}$ progressively increased (i.e., periodic breathing did not occur). During non-REM sleep in a background of hypoxia, both the eupneic levels of $P_{ET_{CO_2}}$ and the threshold for CO_2-dependent apnea were decreased as compared to hyperoxia.

These findings demonstrate the presence of a highly sensitive CO_2-apnea threshold during non-REM sleep. Apneas following spontaneous sighs that transiently decreased $P_{ET_{CO_2}}$ have been reported during light non-REM sleep stages in normal subjects (2). A functional CO_2-apnea threshold may be the critical link between hypoxia-induced periodic breathing and sleep state.

During wakefulness and REM sleep the state-dependent nonchemical influences on inspiratory activity contribute to the maintenance of respiratory cycle rhythm and may prevent periodic breathing and apneas by preventing the expression of a CO_2-apnea threshold.

Mechanism of Hypoxia-Induced Periodic Breathing

According to our findings, the genesis of self-sustaining periodic breathing during non-REM sleep requires the combined influences of both hypoxia and hypocapnia; we suggest that during hypoxia-induced periodic breathing the ventilatory system behaves in a manner consistent with a CO_2-apnea threshold operating within a few mmHg of the eupneic P_{CO_2} obtained during non-REM sleep. Several models explain periodic breathing as the result of instability in the chemical feedback control of ventilation (3–6, 13, 17, 18). That of Cherniack et al. (3, 4) based on observations in anesthetized animals (5, 6) emphasizes the importance of the nonlinear response characteristics and changing gains of the peripheral and medullary chemoreceptors in the production of periodic breathing. Our data from sleeping humans support these concepts.

We propose that the combination of augmented peripheral chemoreceptor gain in hypoxia with the hypocapnia-induced apnea threshold in non-REM sleep can explain both the nature and mechanism of hypoxia-induced periodic breathing according to the following logic (Fig. 11). During the combined conditions of hypoxia and non-REM sleep, increased peripheral chemoreceptor activity lowers P_{CO_2} and $[H^+]$ in arterial plasma and cerebral fluids to near the apnea threshold, thus creating an inherently unstable condition. Ventilation and breathing pattern variability increase as breathing becomes oscilla-

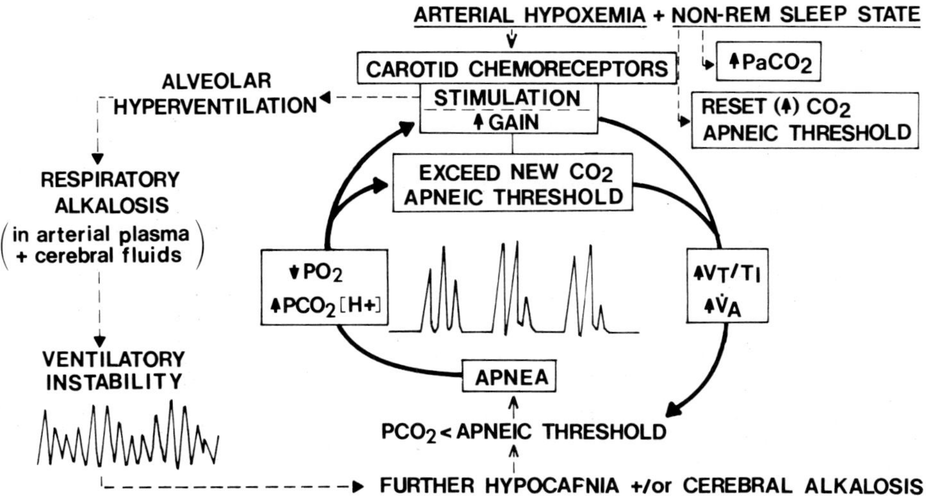

FIG. 11. Schema illustrating a possible mechanism for pathogenesis of hypoxia-induced periodic breathing during non-REM sleep. See text for explanation.

tory, but no apneas appear. A subsequent small drop in Pa_{CO_2} then occurs, possibly because of increasing ventilation, further washout of CO_2 stores secondary to the longer duration of hypoxemia, or spontaneous perturbations in VT secondary to sighs or transient changes in upper airway resistance. Thus Pa_{CO_2} (or more probably cerebral fluid PCO_2 and $[H^+]$) is driven below its threshold, resulting in apnea. During the apnea further hypoxemia develops, which shifts Pa_{O_2} further along that portion of the peripheral chemoreceptor–response curve at which gain rapidly increases. During this period Pa_{CO_2} and $[H^+]$ are also rising and eventually exceed their apnea threshold (i.e., a threshold value lower than that at the start of the apneic period because of the lower background PO_2). After the apnea the augmented chemoreceptor stimuli and gain cause an overshoot, rather than gradual increase, in inspiratory effort (VT/TI) and thus in VT and $\dot{V}E$. This hyperventilation drives Pa_{CO_2} below its (again new) apnea threshold. The cycle then repeats itself, resulting in self-sustained periodic oscillations.

Our model predicts that changes in either the position of the CO_2-apnea threshold or the relative gain of the peripheral chemoreceptors would change the nature of periodic breathing. Altering the net level of ventilatory afferent stimuli has been found to change the position of the CO_2-apnea threshold (19, 26). Thus the stabilizing effects of chronic medroxyprogesterone acetate and acetazolamide administration on periodic breathing observed during sleep at high altitude (29, 31) are probably caused by these background stimuli decreasing the CO_2-apnea threshold or by reduced peripheral chemoreceptor gain secondary to increases in Pa_{O_2}

This work was supported by contracts from United States Army Research and Development Command DAMD 17-77-C-7006, National Institutes of Health Grant HL-15469, and the Medical Research Service of the Veterans Administration. Anne Berssenbrugge was supported by an American Heart Association/Wisconsin Affiliate Fellowship.

REFERENCES

1. BERSSENBRUGGE, A., J. DEMPSEY, C. IBER, J. SKATRUD, AND P. WILSON. Mechanisms of hypoxia-induced periodic breathing during sleep in humans. *J. Physiol. London* 343: 507–524, 1983.

1a. BRUSIL, P. J., T. B. WAGGENER, R. E. KRONAUER, AND P. GULESIAN, JR. Methods for identifying respiratory oscillations disclose altitude effects. *J. Appl. Physiol.: Respirat. Environ. Exercise Physiol.* 48: 545–556, 1980.

2. BULOW, K. Respiration and wakefulness in man. *Acta Physiol. Scand. Suppl.* 209: 1–110, 1963.

3. CHERNIACK, N. S. Respiratory dysrhythmias during sleep. *N. Engl. J. Med.* 305: 325–330, 1981.

4. CHERNIACK, N. S., AND G. S. LONGOBARDO. Cheyne-Stokes breathing—an instability in physiologic control. *N. Engl. J. Med.* 288: 952–957, 1973.

5. CHERNIACK, N. S., G. S. LONGOBARDO, O. R. LEVINE, R. MELLINS, AND A. P. FISHMAN. Periodic breathing in dogs. *J. Appl. Physiol.* 21: 1847–1854, 1966.

6. CHERNIACK, N. S., C. VON EULER, I. HOMMA, AND F. F. KAO. Experimentally induced Cheyne-Stokes breathing. *Respir. Physiol.* 37: 185–200, 1979.

7. DOUGLAS, C. G., AND J. S. HALDANE. The causes of periodic or Cheynes-Stokes breathing. *J. Physiol. London* 38: 401–419, 1909.

8. ENGLAND, S. J., D. BARTLETT, JR., AND S. L. KNUTH. Comparison of human vocal cord movements during isocapnic hypoxia and hypercapnia. *J. Appl. Physiol.: Respirat. Environ. Exercise Physiol.* 53: 81–86, 1982.

9. FINK, B. R. Influence of cerebral activity in wakefulness on regulation of breathing. *J. Appl. Physiol.* 16: 15–20, 1961.

10. FORSTER, H. V., J. A. DEMPSEY, J. THOMSON, E. VIDRUK, AND G. A. DOPICO. Estimation of arterial PO_2, PCO_2, pH, and lactate from arterialized venous blood. *J. Appl. Physiol.* 32: 134–137, 1972.

11. GAUTIER, H., AND M. BONORA. Possible alterations in brain monoamine metabolism during hypoxia-induced tachypnea in cats. *J. Appl. Physiol.: Respirat. Environ. Exercise Physiol.* 49: 769–777, 1980.

12. GOTHE, B., M. D. ALTOSE, M. D. GOLDMAN, AND N. S. CHERNIACK. Effect of quiet sleep on resting and CO_2-stimulated breathing in humans. *J. Appl. Physiol.: Respirat. Environ. Exercise Physiol.* 50: 724–730, 1981.

13. GUYTON, A. C., J. W. CROWELL, AND J. W. MOORE. Basic oscillating mechanism of Cheyne-Stokes breathing. *Am. J. Physiol.* 187: 395–398, 1951.
14. IBER, C., A. BERSSENBRUGGE, J. B. SKATRUD, AND J. A. DEMPSEY. Ventilatory adaptations to resistive loading during wakefulness and non-REM sleep. *J. Appl. Physiol.: Respirat. Environ. Exercise Physiol.* 52: 607–614, 1982.
15. KALAPESI, Z., M. DURAND, F. N. LEAKY, D. B. CATES, M. MACCALLUM, AND H. RIGATTO. Effect of periodic or regular respiratory pattern on the ventilatory response to low inhaled CO_2 in preterm infants during sleep. *Am. Rev. Respir. Dis.* 123: 8–11, 1981.
16. KELLOGG, R. H. Balloons and mountains. *Oxford Med. Sch. Gazette* 23: 15–18, 1971.
17. KHOO, M. C. K., R. E. KRONAUER, K. P. STROHL, AND A. S. SLUTSKY. Factors inducing periodic breathing in humans: a general model. *J. Appl. Physiol.: Respirat. Environ. Exercise Physiol.* 53: 644–659, 1982.
18. MILHORN, H. T., JR., AND A. C. GUYTON. An analog computer analysis of Cheyne-Stokes breathing. *J. Appl. Physiol.* 20: 328–333, 1965.
19. MITCHELL, R. A., C. R. BAINTON, AND G. EDELIST. Posthyperventilation apnea in awake dogs during metabolic acidosis and hypoxia. *J. Appl. Physiol.* 21: 1363–1367, 1966.
20. PHILLIPSON, E. A. Control of breathing during sleep. *Am. Rev. Respir. Dis.* 118: 909–939, 1978.
21. RECHTSCHAFFEN, A., AND A. KALES. *A Manual of Standardized Terminology, Techniques and Scoring System for Sleep Stages of Human Subjects.* Washington, DC: Natl. Inst. Health, Publ. 204, 1968.
22. REITE, M., D. JACKSON, R. L. CAHOON, AND J. V. WEIL. Sleep physiology at high altitude. *Electroencephalogr. Clin. Neurophysiol.* 38: 463–471, 1975.
23. REMMERS, J. E. Effects of sleep on control of breathing. In: *Respiratory Physiology III*, edited by J. G. Widdicombe. Baltimore, MD: University Park, 1981, vol. 23, p. 111–147. (Int. Rev. Physiol. Ser.)
24. REMMERS, J. E., D. BARTLETT, JR., AND M. D. PUTNAM. Changes in the respiratory cycle associated with sleep. *Respir. Physiol.* 28: 227–238, 1976.
25. SANTIAGO, T. V., AND N. H. EDELMAN. Mechanisms of ventilatory response to carbon monoxide. *J. Clin. Invest.* 57: 977–986, 1976.
26. SKATRUD, J., AND J. DEMPSEY. Effect of sleep on ventilatory response to alkalosis in humans (abstr.). *Federation Proc.* 41: 1103, 1982.
27. SKATRUD, J. B., J. A. DEMPSEY, C. IBER, AND A. BERSSENBRUGGE. Correction of CO_2 retention during sleep in patients with chronic obstructive pulmonary disease. *Am. Rev. Respir. Dis.* 124: 260–268, 1981.
28. SULLIVAN, C. E., L. F. KOZAR, E. MURPHY, AND E. A. PHILLIPSON. Primary role of respiratory afferents in sustaining breathing rhythm. *J. Appl. Physiol.: Respirat. Environ. Exercise Physiol.* 45: 11–17, 1978.
29. SUTTON, J. R., C. S. HOUSTON, A. L. MANSELL, M. D. MCFADDEN, P. M. HACKETT, J. R. A. RIGG, AND A. C. P. POWLE. Effect of acetazolamide on hypoxemia during sleep at high altitude. *N. Engl. J. Med.* 301: 1329–1331, 1979.
30. WAGGENER, T. B., P. J. BRUSIL, R. E. KRONAUER, AND R. GABEL. Strength and period of ventilatory oscillations in unacclimatized humans at high altitude (abstr.). *Physiologist* 20: 9, 1977.
31. WEIL, J. V., M. H. KRYGER, AND C. H. SCOGGEN. Sleep and breathing at high altitude. In: *Sleep Apnea Syndromes*, edited by C. Guilleminault and W. C. Dement. New York: Liss, 1978, chapt. 7, p. 119–135.

11

Mechanisms for Recurrent Apneas at Altitude

NEIL S. CHERNIACK, BARBARA GOTHE,
AND KINGMAN P. STROHL

Department of Medicine, Case Western Reserve University, Cleveland, Ohio

BREATHING IS COMMONLY INTERRUPTED by periods of apnea during sleep, particularly at altitude (39). These apneas have recently been classified into two types (5, 20, 28). Central apneas are those in which all respiratory activity disappears for a time; in obstructive apneas respiratory activity is continuous but airflow stops. Temporary failure of the central respiratory-pattern generator is believed to cause central apneas. Upper air blockade, on the other hand, is believed to cause obstructive apneas. Both types of apnea can be sporadic or recurrent. During sleep at sea level, apneas are most common in the very young and in the elderly and occur more frequently in males than in females (20, 28). Such apneas can be central, obstructive, or mixed (where apneas begin as central and continue during an obstructed phase), and all types can be observed in the same individual in a single night's sleep (20, 28).

At altitude, apneas are recurrent and frequently resemble the breathing described by Cheyne and Stokes (Cheyne-Stokes respiration) with a regular waxing and waning of tidal volume and often breathing rate (12, 30). Central apneas appear to be common during sleep at altitude, but the relative frequency of obstructive and central apneas is not known. Recurrent apneas are difficult to explain because it is necessary to account for not only the cessation of breathing or airflow but its recurrent nature. Many theories have been formulated, but none have met with universal acceptance. There are three reasons for this. First, not enough is known of the normal regulation of breathing; second, sleep studies are time consuming and technically much more difficult than similar studies in awake subjects; finally, recurrent apneas (particularly obstructive apneas) are difficult to reproduce in animal models.

RESPIRATORY CONTROL SYSTEM

Because apneas disturb the normal breathing sequence, the mechanisms that regulate breathing must be understood to explain why breathing stops. However, essential information is missing. For example, we still are uncertain of the origin of the respiratory drives that stimulate breathing during rest, exercise, or CO_2 inhalation (44).

Figure 1 diagrams the respiratory system, indicating some of the feedback loops through which changes just in chemical drive could influence breathing. The diagram, however, incompletely represents respiratory control by omitting the modulating action of mechanoreceptors in the upper airways, lung, and chest wall; baroreceptors, which affect both respiration and circulation; and the temperature-regulating system, which can exert a powerful effect on respiration. Also the known drives to breathe interact because of coupling among the various feedback loops that seem to be involved in respiratory regulation. However, the figure highlights some of the features of the system that could account for recurrent apneas.

Like other control systems the respiratory system can be conceptualized as an interconnected controller and a controlled system. The controller consists of the bulbopontine respiratory-pattern generator and the chemoreceptors (peripheral and central) (8, 18, 44). Also part of the controller, and particularly important for understanding apneas, are higher brain centers; they can influence the excitability of the pattern generator and are involved in the sleep-waking cycle (5, 28). The plant or controlled system consists of the lungs, the respiratory muscles, the chest wall skeleton, and the blood and tissues in which the controller is imbedded (7).

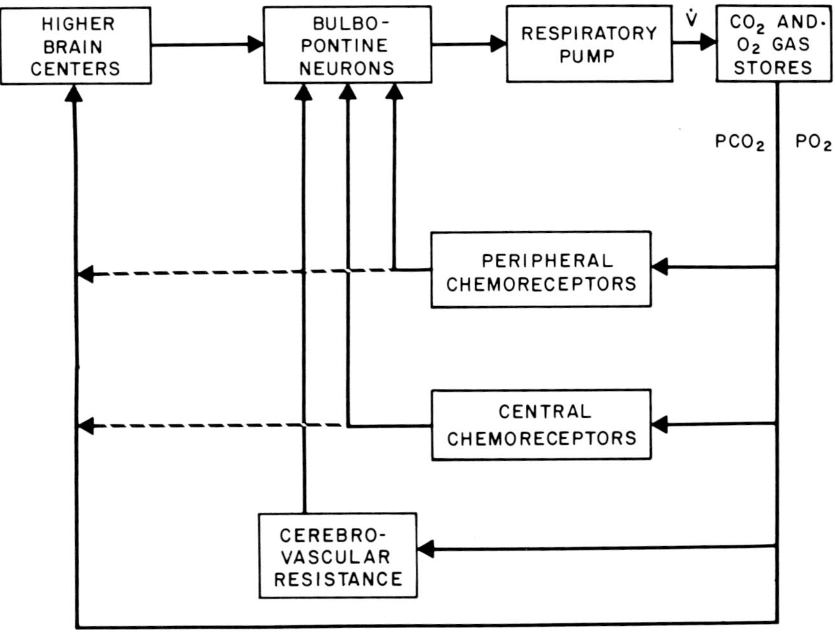

FIG. 1. Block diagram of chemical control of respiration. Partial pressures of CO_2 and O_2 (P_{CO_2} and P_{O_2}) affect respiration by acting on peripheral and central chemoreceptors, cerebral vascular resistance, and higher brain centers. (No direct effect of P_{O_2} on central chemoreceptors.) Action of P_{CO_2} and P_{O_2} on higher brain centers may be direct (*solid lines*) or mediated by peripheral and central chemoreceptors (*broken lines*).

Hypoxia and hypercapnia can influence breathing in several ways. They affect cerebral blood flow and cardiac output. Also they can increase or decrease the activity of higher brain centers (5, 8, 18, 35).

In this kind of system recurrent apneas could occur because the output of the respiratory system is normally oscillatory. Sleep or altitude might exaggerate the normally oscillatory behavior of ventilation (not the normal cycle of inspiration and expiration within a breath but the oscillations in ventilation that occur over a long time). On the other hand, the output of the respiratory system (like other control systems) in the completely relaxed but awake state may be relatively constant, but sleep or altitude change the system in characteristic ways that cause the usually stable behavior of man-made control systems to become oscillatory.

OSCILLATIONS DURING NORMAL BREATHING

Oscillations in the activity of physiological systems are being recognized more and more often (36, 42, 44). Many of them could affect respiration. The normal sequence of waking, quiet, and rapid-eye-movement (REM) sleep exemplifies the activity of a "biological clock," which produces periodic changes in breathing (ventilation increasing during wakefulness and decreasing during sleep).

Oscillation varying in cycle length (periodic breathing) can be observed even in conscious subjects (14, 23). Although apneas are rare no evidence demonstrates major differences in the mechanism that could cause either recurrent apneas or merely cyclic changes in ventilation. These oscillations are more evident with sophisticated filtering techniques, which can separate oscillations with different periods (37). Oscillations in breathing may be partially obscured during wakefulness by the ventilatory excitatory effects of nonspecific stimuli such as light and sound. It has been proposed that a "wakefulness drive" presumably produced by nonspecific environmental stimuli helps prevent apnea in conscious subjects when chemical drives to breathing are reduced (13). During sleep, as the effects on ventilation of nonspecific stimulation subside, the oscillations already present in the waking state may become more apparent (Fig. 2). New rhythms may also develop during sleep,

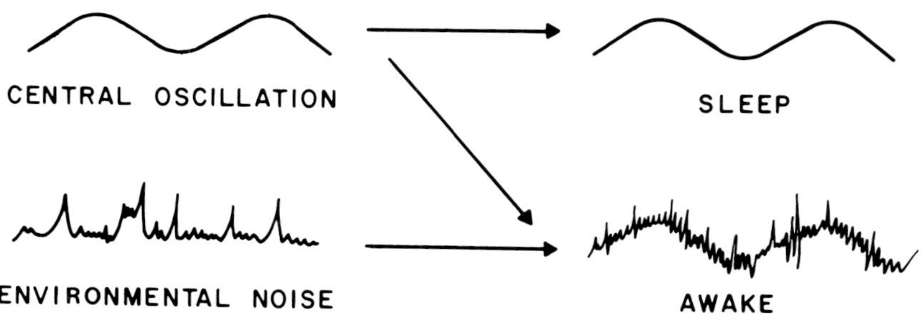

FIG. 2. Diagram indicating that a normally present central oscillation in ventilation may be obscured by a nonspecific environmental noise (light, sound, etc.) during wakefulness and be obvious during sleep when these nonspecific effects on ventilation disappear.

accentuating the cycling of breathing. The repetitive pontogeniculo-occipital waves observed during REM sleep may exemplify new recurrent phenomena that drive breathing cyclically during somnolence (28).

Chemoreceptors might be expected to smooth such ventilatory oscillations. For example, the hypoxia and hypercapnia developing during the nadir of a ventilatory oscillation would abbreviate apnea by stimulating chemoreceptors. However, if chemosensitivity was blunted or if hypoxic depression developed, apneas might be more prolonged and hence more evident (6, 31).

Apparently irregular breathing while awake is more common in subjects with primary alveolar hypoventilation (43). Periodic breathing during sleep has been reported in such individuals and in dogs after peripheral chemoreceptor denervation (28). Hypoxic depression developing during apnea (which would be intensified at altitude) might potentiate cycles by further reducing the effect of chemoreceptor activation (6).

However, no clear relationship has been demonstrated between chemosensitivity and repeated apneas. Although patients with the sleep-apnea syndrome may have depressed ventilatory responses to chemical stimuli while awake, their chemosensitivity seems to improve if treatments eliminate the apneas (21). This suggests that the reduced responses to chemical stimuli observed during wakefulness may result from the sleep apnea rather than cause it. Also, although treatment of sleep apnea with respiratory stimulants has been dramatically successful sometimes, it seems to have had little beneficial effect in other instances (34). Nonetheless the periodicities observed in normal breathing probably have some impact on sleep apneas.

OSCILLATIONS AND UPPER AIRWAY MUSCLES

When the inspiratory chest wall muscles contract, they create a negative pressure in the upper airway that tends to cause collapse (1, 2, 30). Structural abnormalities that narrow the upper airways or increase the pressure drop also increase the probability of airway blockade. Collapsing forces seem to be resisted by the mechanical properties of the upper airway muscles like the posterior cricoarytenoid, genioglossus, and ala nasi, which dilate the laryngeal, oropharyngeal, and nasal passages (1, 28, 30, 33, 40). During sleep the activity of these muscles decreases more than the activity of the diaphragm so that obstruction is more likely. The recumbent posture assumed during sleep also enhances gravitational effects that tend to produce upper airway closure. Decreased activity of the nasal, pharyngeal, and laryngeal muscles (muscles inserting on the hyoid) have each been thought crucial in the production of obstructive apneas (17, 28, 30, 36). The convoluted structure of the upper airway passages and the potential complex interplay of various muscles in producing pressure and flow changes suggest that the relative importance of the various muscles in obstructive apnea may vary considerably between individuals.

It has been suggested that the relationship between onset times of upper airway muscle and chest wall muscles also affects airway patency (33). In conscious subjects the activity of muscles like the ala nasi and the genioglossus precedes the onset of diaphragm electrical activity (26, 33). In sleep-apnea patients the activity of the two sets of muscles can be coincident or upper airway muscle activity may begin after the commencement of diaphragm contraction. Because the dilator upper airway muscles contract too late and the negative pressure in the airway produced initially by the descent of the diaphragm is unopposed, obstruction that cannot be subsequently reversed during that inspiratory effort may occur.

This idea can be extended. Oscillations both in upper airway muscle activity and diaphragm activity may be present normally. These two kinds of oscillations need not be in phase, as shown in Figure 3, and may produce either mixed or obstructive apneas, depending on the phase difference (5, 9). We have observed cyclic changes in genioglossus activity in the unrestrained sleeping cat (M. Haxhiu, personal communication). Although obstruction was not evident, more pronounced oscillations might have prevented airflow.

Like the diaphragm the upper airway muscles increase their activity with chemical stimulation (1, 5, 26, 27). Sleep diminishes the effect of chemical stimuli on the phasic and tonic activity of the upper airway muscles and hence would be likely to reduce any smoothing action of chemical feedback on oscillatory behavior (28). In addition the responses of the upper airway muscles to chemical stimuli differ from those of the chest wall muscles. For example, in anesthetized cats hypoglossal nerve activity seems to increase more steeply than phrenic nerve activity at high levels of chemical drive and less steeply at low levels of chemical drive (40). The activity of the sternohyoid muscle seems to plateau with hypercapnia before that of the diaphragm (38). In anesthetized animals both these muscles seem to require more hypoxia and hypercapnia than the diaphragm for activity to begin (1, 27, 40). In other words they seem to have higher thresholds.

The dynamic responses of the upper airway and chest wall muscles can also differ. For example, we have recently observed that the electrical activity of the genioglossus increases more slowly than that of the diaphragm in awake or sleeping cats made to inspire CO_2. On the other hand, genioglossus electrical

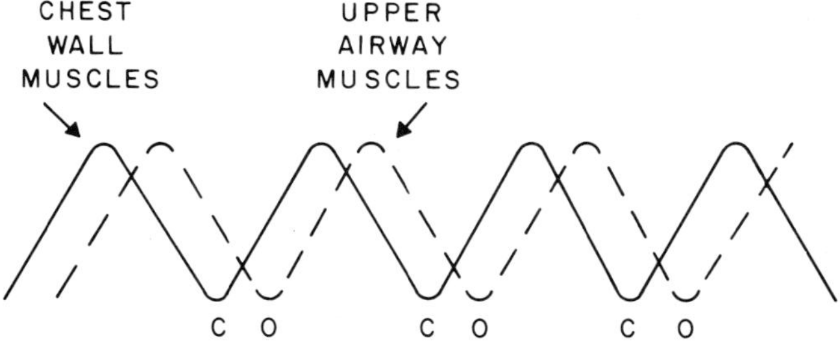

FIG. 3. Diagram illustrating effect of possible out-of-phase oscillations in chest wall and upper airway muscle activity. When chest wall muscle activity reaches its nadir, central apnea (C) occurs; when upper airway muscle activity reaches its nadir, obstructive apnea (O) occurs.

activity decreases more rapidly when the CO_2 is removed from the inspired gas. These dissimilarities in the behavior of upper airway and chest wall muscles could contribute to phase shifts in activity and differences in amplitude response even if both sets of muscles responded to a common oscillatory input. Thus they can produce mixed or obstructive apneas.

An increase in the activity of constrictor muscles in the upper airways rather than a decrease in dilator muscle activity might cause obstructive apneas. While these constrictors have been studied much less than the dilators, cyclic changes in their output may also contribute to obstructive apnea.

Neurological abnormalities also may intensify oscillations in breathing (3, 12). Recurrent apneas are a reported characteristic of the ataxic breathing observed in patients with bulbopontine lesions (3). Webber has produced Biot's breathing (clusters of breaths with relatively fixed tidal volumes separated by apneas) in cats with pneumotaxic center lesions after careful administration of pentobarbital (38). Too much pentobarbital causes gasping breathing, while too little results in apneustic breathing. Chemoreceptor input in this type of preparation has the predictable smoothing effect. Hypoxia and hypercapnia shorten the apneic period, but hyperoxia and carotid sinus denervation seem to lengthen it.

INSTABILITY IN FEEDBACK CONTROL AS A CAUSE OF RECURRENT APNEAS

Regular oscillations in breathing could develop during sleep even if the main cause of breathing variations in the awake state was random input from environmental noise. Neither is a neurological lesion necessary to produce recurrent apneas.

It is well known that man-made control systems can become unstable in predictable circumstances (7, 5, 9). The instability is manifested by the conversion of a usually stable output to a cyclic one. The changes that produce it include delays in transmission of information around the system, increases in controller sensitivity (gain), and reduction in system damping (5, 8).

In the respiratory system prolonging circulation time can lengthen delays so that the interval required for chemoreceptors to sense changes in gas tensions in the alveoli is increased. The interaction of hypoxia with the ventilatory response to CO_2 may increase controller gain. Damping can be altered by changing the properties of the gas stores in the body that affect the speed with which blood gas tensions respond to changes in ventilation. Hypoxia caused by reducing the already small amount of O_2 contained in the lungs and in the blood reduces damping. All these changes can be demonstrated experimentally and through equations but unfortunately are difficult to explain with words. Essentially these changes result in instability because the controller is informed too late of its effects on the partial pressures of CO_2 and O_2 (P_{CO_2} and P_{O_2}, respectively).

For example, the chemoreceptors become sensitive to changes in lung gas tensions through the circulation. If the controller increases ventilation to lower arterial P_{CO_2} while the circulation time is prolonged, increased ventilation may continue too long. This could drive arterial P_{CO_2} below the threshold level needed to maintain ventilation. In the absence of a wakefulness drive, apnea results (13). As P_{CO_2} rises and P_{O_2} falls during apneas, breathing resumes. If the circulation time is still lengthened, however, apnea is maintained too long. When breathing resumes it is excessive because of the interaction of hypoxia with CO_2 and the cycle continues. Periodic breathing of the Cheyne-Stokes type is common in patients with congestive heart failure and long circulation times (8). Periodic breathing has also been produced in dogs by lengthening the circulation time (22).

The hypoxia of altitude increases the tendency of the respiratory control system to become unstable in several ways. Although ventilation increases with hypercapnia are relatively linear, the ventilatory response to hypoxia increases as hypoxia becomes more severe (Fig. 4). Hence the controller gain ($\Delta\dot{V}/\Delta P_{O_2}$, where $\dot{V}$ is ventilation) increases as P_{O_2} is decreased (5). Moreover the effect of hypercapnia in increasing the ventilatory response is intensified as arterial P_{O_2} falls. Increased ventilatory response increases the speed of gas tension changes so that information may need to be transferred more quickly to prevent instability.

Hypoxia caused by lowering the total amount of O_2 in the body heightens the changes in arterial P_{O_2} produced by ventilation changes (5, 8). This is further exaggerated in the blood O_2 stores by the shape of the oxyhemoglobin-dissociation curve (7). The functional residual capacity of the lung decreases in the recumbent position adopted during sleep, which reduces the O_2 stores in the lung (7, 5). On the other hand, the larger CO_2 stores and the relatively lower CO_2-dissociation curve tend to promote instability. Sleep may potentiate these destabilizing effects of hypoxia in several ways: by eliminating nonspe-

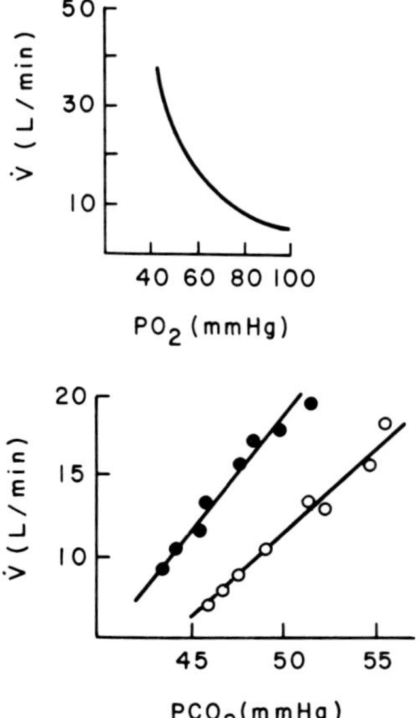

FIG. 4. *Top*: ventilatory response to hypoxia awake. Because of the curvilinearity of response at low levels of P_{O_2}, controller gain ($\Delta\dot{V}/\Delta P_{O_2}$) is greater than at higher levels of P_{O_2}. Response to P_{O_2} asleep is similarly curved but may be somewhat flatter. *Bottom*: ventilatory response to hypercapnia awake (●) and asleep (○). Responses to CO_2 are linear. With sleep, however, response line is shifted to the right, indicating that higher levels of P_{CO_2} are needed to produce a given level of ventilation and suggesting that higher levels of P_{CO_2} may be required to initiate breathing during sleep.

cific drives to ventilation that prevent apnea, by increasing the threshold level of P_{CO_2} and P_{O_2} needed for breathing to resume after apnea so that periods where ventilation is absent are greater, or by lengthening circulation time.

These explanations for periodic breathing at altitude are consistent with the following observations. *1*) Periodic breathing occurs in anesthetized dogs after a period of apnea (produced by hyperventilation) to hypocapnia (10). *2*) Periodic breathing also occurs in sleeping humans after hypocapnia and apnea are produced by hyperventilation (J. Dempsey, personal communication). *3*) In cats periodic breathing can be produced by hypoxia and relieved by O_2 (11). *4*) In humans O_2 breathing diminishes or terminates the frequency of central apneas (16). *5*) Finally, periodic breathing at altitude is less common in native highlanders who have diminished hypoxic drives (S. Lahiri, personal communication).

More difficult to explain is the observation by Lahiri that extreme hypoxia and hyperoxia diminish recurrent apneas in sleeping newcomers to altitude. Possibly the increased breathing with extreme hypoxia is similar to the tachypnea reported by Edelman (31) in the late stages of hypoxic depression caused by CO inhalation. Another possibility is that at low levels of hypoxia a "dogleg" develops where ventilation is unaffected by changes in P_{CO_2} (similar to those at all levels of hypoxia in awake state).

In other feedback loops changes affecting respiration could also contribute to the instabilities observed during sleep. Bulow (4) reported that periodic apneas occurred more often in subjects who appeared to increase their P_{CO_2} threshold most during sleep. An apparent shift in the CO_2 threshold has been noted in many studies measuring CO_2 sensitivity during sleep [Fig. 4; (15)]. Phillipson (28) used this shift to explain periodic breathing during sleep. He suggested that this shift produces apnea at sleep onset. The hypoxia and hypercapnia that develop during the apnea act on the brain to produce arousal. The awakening increases breathing and lowers gas tensions so that sleep recurs.

Although it is still uncertain whether the actions of hypoxia and hypercapnia on awakening are direct effects or mediated through chemoreceptors, clearly both sufficiently intense hypoxia and hypercapnia can arouse sleeping subjects. This arousing effect varies from subject to subject and may be affected by previous sleep history. Results by B. Gothe (personal communication) indicate that hypoxia and hypercapnia potentiate each other's effects on arousal, but the quantitative relationship between hypoxia and hypercapnia as arousing stimuli remains to be determined. Neither has experimental evidence indicated a relationship between sensitivity to arousal by hypoxia or hypercapnia and the occurrence of sleep apnea. Nevertheless this kind of mechanism would contribute to instability in the control system. Decreased arousal responses might also explain the more regular breathing usually observed in the deeper stages of quiet sleep but so would a generalized decrease in ventilatory responses to all exogenous and endogenous stimuli (28).

Hypercapnia and hypoxia increase cerebral blood flow. This could be a stabilizing factor in respiratory control by diminishing the fluctuations in

P_{CO_2} and H^+ occurring at the central chemoreceptors. Recently Santiago et al. (32) reported large increases in cerebral blood flow during REM sleep and an enhanced cerebral blood flow response to CO_2. The increases in cerebral blood flow are disproportionate to any changes in cerebral production of CO_2 during REM sleep so that brain tissue levels of P_{CO_2} and H^+ should be decreased. They have suggested that these changes in cerebral blood flow may be sufficient to bring P_{CO_2} below the level of H^+ needed to excite central chemoreceptors and thus produce apnea. The hypoxia of altitude could potentiate these cerebral blood flow effects. Although the contribution of brain blood flow to sleep apnea is still unclear, it could also contribute to instability in respiratory control. It is difficult to know which of the mechanisms described contributes most to sleep apneas. In practice the effects of changes in P_{CO_2} and P_{O_2} on all the feedback loops indicated in Figure 1 would determine the stability or instability of breathing.

Because the instability produces cyclic changes in blood gas tensions, the activity of the upper airway muscles like that of the chest wall muscles will cycle. The different effects of sleep and chemical stimuli on upper airway and chest wall muscles (outlined in OSCILLATIONS AND UPPER AIRWAY MUSCLES, p. 132) cause blood gas cycles to produce obstructive and mixed apnea. The nonlinearity of the response of the upper airway muscles to chemical stimuli (if they behave like hypoglossal nerves in anesthetized cats) could account for the periodic relief of the obstructive apneas (Fig. 5; 25, 27, 40). However, studies in humans suggest that upper airway responses to hypercapnia and hypoxia are more linear (26). In humans the arousal effects (e.g., of CO_2 and hypoxia) on upper airway muscles seem to be greater than on the chest wall muscles, which could contribute to an apparent nonlinearity of the response of upper airway muscles to chemical drive (30). It has been suggested that in humans even small changes in arousability, perhaps undetectable in the electroencephalogram, might greatly change the activity of the muscle in the upper airways (28).

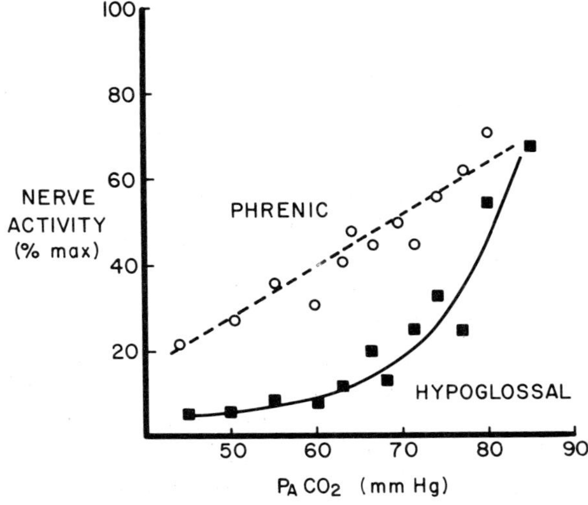

FIG. 5. Effect of CO_2 on phrenic and hypoglossal activity in an anesthetized dog. Note nonlinear response of hypoglossal nerve. [From Weiner et al. (40).]

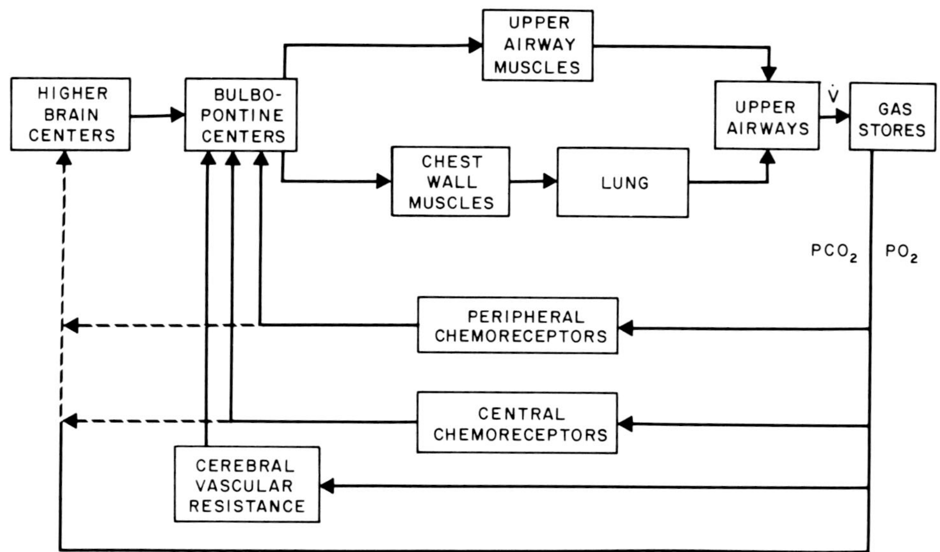

FIG. 6. Model of chemical control of ventilation, which indicates upper airway muscles. Upper airway resistance depends on activity of both chest wall and upper airway muscles and hence on their relative responses to chemical stimuli.

Figure 6, a block diagram that includes the effects of chemical stimuli on upper airway muscles and chest wall muscles, like Figure 1, omits the effects of mechanoreceptors on breathing. Mechanoreceptors modify the pattern and intensity of chest wall muscle activity and seem to have even greater effects on upper airway muscles (25, 27). Airway obstruction in anesthetized animals seems to increase the activity of the airway dilators more than the chest wall muscles by exciting both pulmonary and upper airway receptors. These mechanoreceptor responses and arousal could disproportionally increase dilator muscle activity as obstructive apnea progresses. Apnea, on the other hand, can sometimes be produced by laryngeal stimulation (35).

Oscillations in blood pressure can cause oscillations in respiration (8, 29). Phrenic nerve activity is decreased when blood pressure is elevated by baroreceptor stimulation, but the activity of the genioglossus is reduced even more (19, 41). Hence blood pressure fluctuations might cause either central or obstructive apnea.

Signals from proprioreceptors can also affect ventilation. It has been suggested that in infants apnea can be produced by inhibitory signals sent to the controller from fatiguing respiratory muscles (24).

The two different ways recurrent apneas could be produced during sleep at altitude that we have presented may not be mutually exclusive. Rhythmic oscillations in breathing may be intensified during sleep, but breathing may develop an additional oscillation because of conditions promoting instability in respiratory control during sleep. In some situations (e.g., when chemosensitivity is greatly suppressed) oscillations usually present in breathing may become more apparent. On the other hand, changes in chemosensitivity that act on ventilation, cerebral blood flow, and arousal during sleep may produce an instability in control and thereby a new oscillation that dominates the breathing pattern. It may be that both excessive and too little chemical drive can produce recurrent apnea. The real question may not be why apneas ever occur during sleep but why they occur so infrequently.

REFERENCES

1. BROUILLETTE, R. T., AND B. T. THACH. A neuromuscular mechanism maintaining extrathoracic airway patency. *J. Appl. Physiol.: Respirat. Environ. Exercise Physiol.* 46: 772–779, 1979.
2. BROUILLETTE, R. T., AND B. T. THACH. Control of genioglossus muscle inspiratory activity. *J. Appl. Physiol.: Respirat. Environ. Exercise Physiol.* 49: 801–808, 1980.
3. BROWN, H. W., AND F. PLUM. The neurologic basis of Cheyne-Stokes respiration. *Am. J. Med.* 30: 849–860, 1961.
4. BULOW, K. Respiration and wakefulness in man. *Acta Physiol. Scand.* 59: 1–110, 1963.
5. CHERNIACK, N. S. Respiratory dysrhythmias in sleep. *N. Engl. J. Med.* 305: 325–330, 1981.
6. CHERNIACK, N. S., N. H. EDELMAN, AND S. LAHIRI. Hypoxia and hypercapnia as respiratory stimulants and depressants. *Respir. Physiol.* 11: 113–126, 1970.
7. CHERNIACK, N. S., AND G. S. LONGOBARDO. Oxygen and carbon dioxide gas stores of the body. *Physiol. Rev.* 50: 196–243, 1970.
8. CHERNIACK, N. S., AND G. S. LONGOBARDO. Cheyne-Stokes breathing. *N. Engl. J. Med.* 288: 952–957, 1973.
9. CHERNIACK, N. S., G. S. LONGOBARDO, B. GOTHE, AND D. WEINER. Interactive effects of central and obstructive apnea. In: *Respiration*, edited by I. Hutas and A. Debreczeni. New York: Pergamon, 1981, vol. 10, p. 553–560. (Adv. Physiol. Sci. Ser.)
10. CHERNIACK, N. S., G. S. LONGOBARDO, O. R. LEVINE, R. MELLINS, AND A. P. FISHMAN. Periodic breathing in dogs. *J. Appl. Physiol.* 21: 1847–1854, 1966.
11. CHERNIACK, N. S., C. VON EULER, I. HOMMA, AND F. F. KAO. Experimentally induced Cheyne-Stokes breathing. *Respir. Physiol.* 37: 185–200, 1979.
12. DOWELL, A. R., C. E. BUCKLEY, R. COHEN, R. E. WHALEN, AND H. O. SIEKER. Cheyne-Stokes respiration: a review of clinical manifestations and critique of physiological mechanisms. *Arch. Intern. Med.* 127: 712–726, 1971.
13. FINK, B. R. Influence of cerebral activity in wakefulness on regulation of breathing. *J. Appl. Physiol.* 16: 15–20, 1961.
14. GOODMAN, L., D. M. ALEXANDER, AND D. G. FLEMING. Oscillatory behavior of respiratory gas exchange in resting man. *IEEE Trans. Bio. Med. Eng.* 13: 57–64, 1966.
15. GOTHE, B., M. D. ALTOSE, M. D. GOLDMAN, AND N. S. CHERNIACK. Effect of quiet sleep on resting and CO_2-stimulated breathing in humans. *J. Appl. Physiol.: Respirat. Environ. Exercise Physiol.* 50: 724–730, 1981.
16. GOTHE, B., G. S. LONGOBARDO, P. MANTEY, M. D. GOLDMAN, AND N. S. CHERNIACK. Effect of increased inspired oxygen and carbon dioxide on periodic breathing during sleep. *Trans. Assoc. Am. Physicians* 94: 134–143, 1981.
17. GOTTFRIED, S. B., K. P. STROHL, A. F. DIMARCO, AND W. VAN DE GRAAFF. Upper airway resistance during phrenic stimulation in anesthetized dogs. *Am. Rev. Respir. Dis.* 125: 219, 1982.
18. GRODINS, F. S., J. BUELL, AND A. J. BART. Mathematical analysis and digital simulation of the respiratory control system. *J. Appl. Physiol.* 22: 260–276, 1967.
19. GRUNSTEIN, M. M., J. P. DERENNE, AND J. MILIC-EMILI. Control of depth and frequency of breathing during baroreceptor stimulation in cat. *J. Appl. Physiol.* 39: 395–404, 1975.
20. GUILLEMINAULT, C., J. VAN DEN HOED, AND M. M. MITLER. *Clinical Overview of the Sleep Apnea Syndrome.* New York: Liss, 1978, p. 1–12.
21. GUILLEMINAULT, C., AND J. CUMMISKEY. Progressive improvement of apnea index and ventilatory response to CO_2 after tracheostomy in obstructive sleep apnea syndrome. *Am. Rev. Respir. Dis.* 126: 14–20, 1982.
22. GUYTON, A. C., J. W. CROWELL, AND J. W. MOORE. Basic oscillating mechanism of Cheyne-Stokes breathing. *Am. J. Physiol.* 187: 395–398, 1968.
23. LENFANT, C. Time-dependent variations of pulmonary gas exchange in normal man at rest. *J. Appl. Physiol.* 22: 675–684, 1967.
24. LOPES, J. M., N. L. MULLER, M. H. BRYAN, AND A. C. BRYAN. Synergistic behavior of inspiratory muscles after diaphragmatic fatigue in the newborn. *J. Appl. Physiol.: Respirat. Environ. Exercise Physiol.* 51: 547–551, 1981.
25. MATHEW, O. P., Y. K. ABU-OSBA, AND B. T. THACH. Influence of upper airway pressure changes on genioglossus muscle respiratory activity. *J. Appl. Physiol.: Respirat. Environ. Exercise Physiol.* 52: 438–444, 1982.
26. ÖNAL, E., M. LOPATA, AND T. D. O'CONNOR. Diaphragmatic and genioglossal electromyogram responses to CO_2 rebreathing in humans. *J. Appl. Physiol.: Respirat. Environ. Exercise Physiol.* 50: 1052–1055, 1981.
27. PARKER, D., K. STROHL, J. MITRA, J. SALAMONE, AND N. S. CHERNIACK. Effect of CO_2 and tidal volume in upper airway muscles (abstr.). *Federation Proc.* 40: 481, 1981.
28. PHILLIPSON, E. A. Control of breathing during sleep. *Am. Rev. Respir. Dis.* 118: 904–934, 1978.
29. PREISS, G., S. ISCOE, AND C. POLOSA. Analysis of a periodic breathing pattern associated with Mayer waves. *Am. J. Physiol.* 228: 768–774, 1975.
30. REMMERS, J. E., W. J. DEGROOT, E. K. SAUERLAND, AND A. M. ANCH. Pathogenesis of upper airway occlusion during sleep. *J. Appl. Physiol.: Respirat. Environ. Exercise Physiol.* 44: 931–938, 1978.
31. SANTIAGO, T. V., AND N. H. EDELMAN. Mechanism of the ventilatory response to carbon monoxide (abstr.). *J. Clin. Invest.* 57: 977–986, 1976.
32. SANTIAGO, T. V., E. GUERRA, A. K. SINHA, AND N. H. EDELMAN. Brain blood flow and ventilation during euoxic and hypoxic sleep. *Clin. Res.* 28: 431A, 1980.
33. STROHL, K. P., M. J. HENSLEY, M. HALLETT, N. A. SAUNDERS, AND R. H. INGRAM, JR. Activation of upper airway muscles before onset of inspiration in normal humans. *J. Appl. Physiol.: Respirat. Environ. Exercise Physiol.* 49: 638–642, 1980.
34. STROHL, K. P., M. J. HENSLEY, N. A. SAUNDERS, S.

M. SCHARF, R. BROWN, AND R. H. INGRAM, JR. Progesterone administration and progressive sleep apnea. *J. Am. Med. Assoc.* 245: 1230–1232, 1981.

35. SULLIVAN, C. E., E. MURPHY, L. F. KOZAR, AND E. A. PHILLIPSON. Waking and ventilatory responses to laryngeal stimulation in sleeping dogs. *J. Appl. Physiol.: Respirat. Environ. Exercise Physiol.* 45: 681–689, 1978.

36. VAN DE GRAAFF, W. B., J. MITRA, K. P. STROHL, J. SALAMONE, AND N. S CHERNIACK. Respiratory activity and reflexes of hyoid muscles in the dog (abstr.). *Federation Proc.* 41: 1507, 1982.

37. WAGGENER, T. B., I. D. FRANTZ III, A. R. STARK, AND R. E. KRONAUER. Oscillatory breathing patterns leading to apneic spells in infants. *J. Appl. Physiol.: Respirat. Environ. Exercise Physiol.* 52: 1288–1295, 1982.

38. WEBBER, C. L., AND C. N. PEISS. Pentobarbital induced apneuisis in intact vagotomized and pneumotaxic lesioned cats. *Respir. Physiol.* 38: 37–57, 1979.

39. WEIL, J. V., M. H. KRYGER, AND C. H. SCOGGIN. Sleep and breathing at high altitude. In: *Sleep Apnea Syndromes*, edited by C. Guilleminault and W. C. Dement. New York: Liss, 1978, p. 119–135.

40. WEINER, D., J. MITRA, J. SALAMONE, AND N. S. CHERNIACK. Effect of chemical stimuli on nerves supplying upper airway muscles. *J. Appl. Physiol.: Respirat. Environ. Exercise Physiol.* 52: 530–536, 1982.

41. WEINER, D., D. PARKER, J. MITRA, J. SALAMONE, AND N. S. CHERNIACK. Effect of blood pressure changes on hypoglossal, recurrent laryngeal and phrenic nerve activity. *Am. Rev. Respir. Dis.* 123: 182, 1981.

42. WEVER, R. A. *The Circadian System of Man: Results of Experiments Under Temporal Isolation.* New York: Springer-Verlag, 1979.

43. WOLKOVE, N., M. D. ALTOSE, S. G. KELSEN, AND N. S. CHERNIACK. Respiratory control abnormalities in alveolar hypoventilation. *Am. Rev. Respir. Dis.* 122: 163–167, 1980.

44. YAMAMOTO, W. S. Computer simulation of ventilatory control by both neural and humoral CO_2 signals. *Am. J. Physiol.* 238 (*Regulatory Integrative Comp. Physiol.* 7): R28–R35, 1980.

45. YATES, F. E. Thermodynamics and life. *Am. J. Physiol.* 234 (*Regulatory Integrative Comp. Physiol.* 3): R81–R83, 1978.

12

Effects of Acclimatization on Sleep Hypoxemia at Altitude

JOHN R. SUTTON, GARY W. GRAY, CHARLES S. HOUSTON, AND A. C. PETER POWLES

McMaster University Medical Centre, Hamilton, Ontario, Canada

THE ARTERIAL HYPOXEMIA that occurs in all subjects acutely exposed to high altitude lessens over succeeding days and weeks through ventilatory acclimatization (2). If acclimatization proceeds normally, the partial pressure of oxygen (Po_2) in arterial blood increases over time. If acute mountain sickness (AMS) intervenes and becomes severe, however, acclimatization does not occur and arterial oxygenation worsens (9). There have been many studies on the effect of ventilatory acclimatization, but these have been concerned primarily with the awake state. At altitude the arterial oxygenation is worse during sleep than during the awake state (10, 11, 13), and the degree of hypoxemia during sleep is probably important in determining the ability to acclimatize. Furthermore sleep hypoxemia might be expected to improve as part of the ventilatory acclimatization. However, we did not find such increases in arterial oxygen saturation during sleep in a cross-sectional study comparing groups of subjects who had been at altitude (5,360 m) for varying lengths of time and therefore had different degrees of acclimatization (5, 10). We wondered if any lack of improvement in arterial oxygen saturation with acclimatization had been obscured by the wide ranges in sleep arterial oxygen saturation observed in all groups. Therefore the present longitudinal study, designed to examine the effects of acclimatization on sleep hypoxemia, uses each subject as his or her own control to eliminate differences between individuals.

SUBJECTS AND METHODS

Studies were conducted on five normal healthy subjects (3 males, 2 females), age 22–36 yr. The subjects had spent 3–5 days at a staging camp at 3,290 m, and three of them then climbed to the high camp (5,360 m) over an 8-day period (group I). The two subjects in group III were airlifted from the staging camp to the high camp.

Two sleep studies were conducted at the high camp, the first 1–5 days after arrival and the second 5–10 days later. The subjects in group I had no symptoms at altitude, whereas the subjects in group II had mild symptoms of AMS (morning headache and anorexia) when studied on the first occasion.

The studies were conducted in heated tents. An observer was awake throughout the night. Arterial oxygen saturation was measured continuously with a fiber-optic ear oximeter (Hewlett-Packard). This oximeter had been checked for accuracy against either radial artery or arterialized earlobe capillary blood (3) with an oxygen saturation meter (Radiometer OSM 1) during progressive hypoxia at sea level (7) or during sleep or exercise at altitude (12). Arterial P_{O_2}, P_{CO_2}, and pH were measured on an Instrumentation Laboratory blood-gas analyzer, calibrated with gases previously analyzed with the Lloyd-Haldane apparatus.

RESULTS

All subjects experienced marked sleep hypoxemia during the night, often associated with periodic breathing or apnea. For some the periodic breathing persisted for most of the night. The individual traces are plotted in two formats. Figures 1*A* and 2*A* show the arterial oxygen saturation in absolute terms throughout the night in the first study and after acclimatization. Figures 1*B* and 2*B* replot the same information in terms of percentage cumulative time with the Slutsky and Strohl method for quantifying arterial oxygen saturation during episodic hypoxemia (8). Figure 1 represents the subjects who climbed from the staging camp and were thus better acclimatized. These subjects were studied 2–5 days after reaching the high camp. They spent 50% of the night on the first occasion at an arterial oxygen saturation between 68% and 70%. The second study on these subjects was conducted 5–10 days after the first. On this occasion they spent 50% of the night at arterial saturations between 67% and 78%.

The two subjects who were airlifted to the high camp were studied 1 and 3 days after arrival. They showed great variation in sleep hypoxemia. One subject spent 50% of the night at an arterial oxygen saturation of 64%, whereas the other had profound hypoxemia and spent 50% of the night at an arterial oxygen saturation of 46%. Both had mild AMS. The second study on these subjects was conducted 11 and 8 days after the first study. The subjects now had comparable levels of arterial oxygen saturation, spending 50% of the night at 72% and 73% arterial oxygen saturation with no symptoms (Table 1).

When subjected to statistical analysis the mean saturation on arrival in the five subjects for 50% of the night was 63 ± 9.8% (mean percent ± SD) and after acclimatization 73 ± 5.0% ($0.1 > P > 0.05$). Although the difference is not statistically significant considering the few subjects and the varying conditions of the study, to conclude that acclimatization does not improve arterial oxygen saturation during sleep is almost certainly a type II, or β-, error. The 95% confidence levels on these results suggest that the true change in arterial oxygen saturation during sleep after acclimatization lies between −3.67% and 23.7%. This range includes a clinically significant change. Using

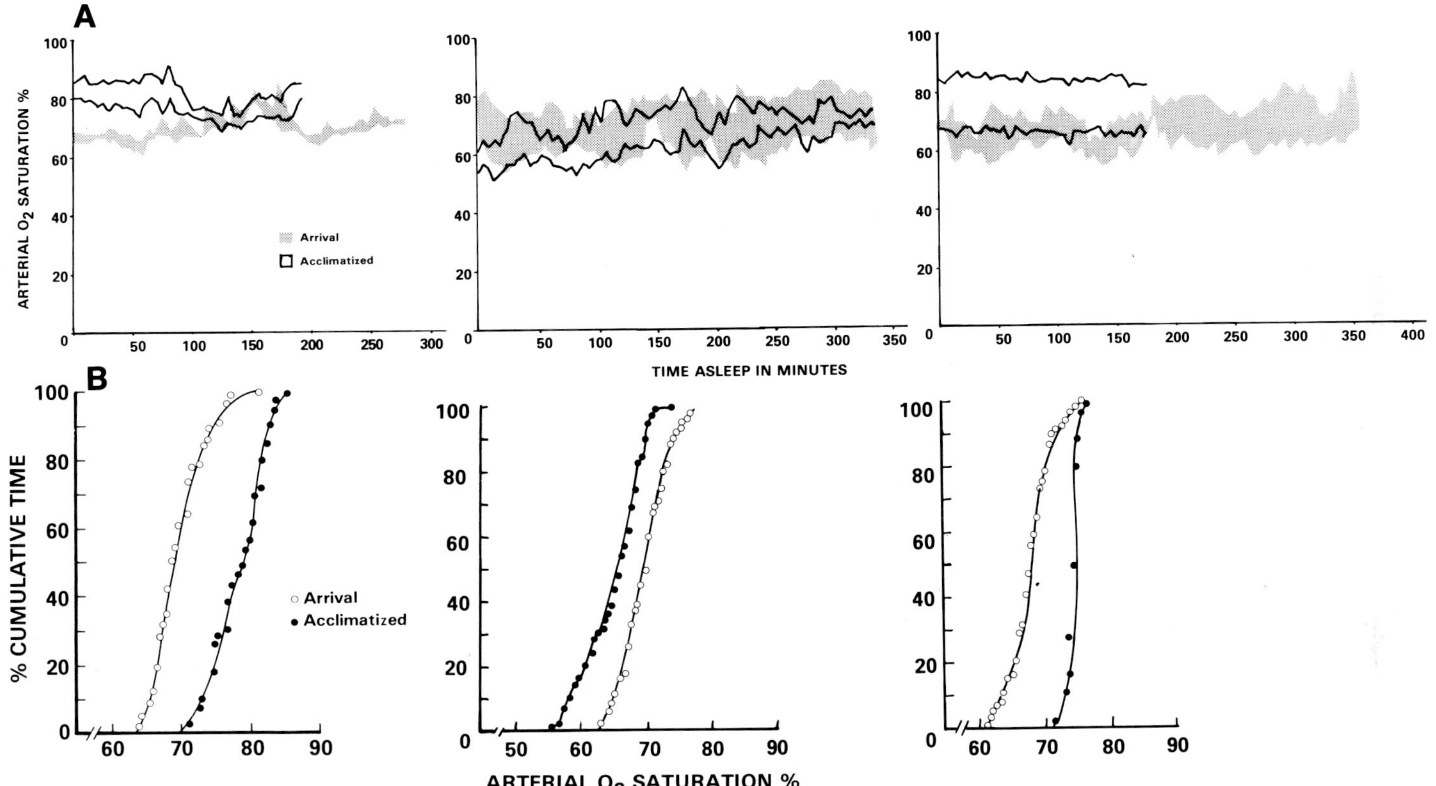

FIG. 1. Sleep saturation at 5,360 m. Subjects in group I (2 females, 1 male) studied 2–5 days after climbing to 5,360 m and again 5–10 days later. *A*: absolute arterial O_2 saturation values. *B*: results replotted with Slutsky and Strohl method for % cumulative time. [From Powles and Sutton (6).]

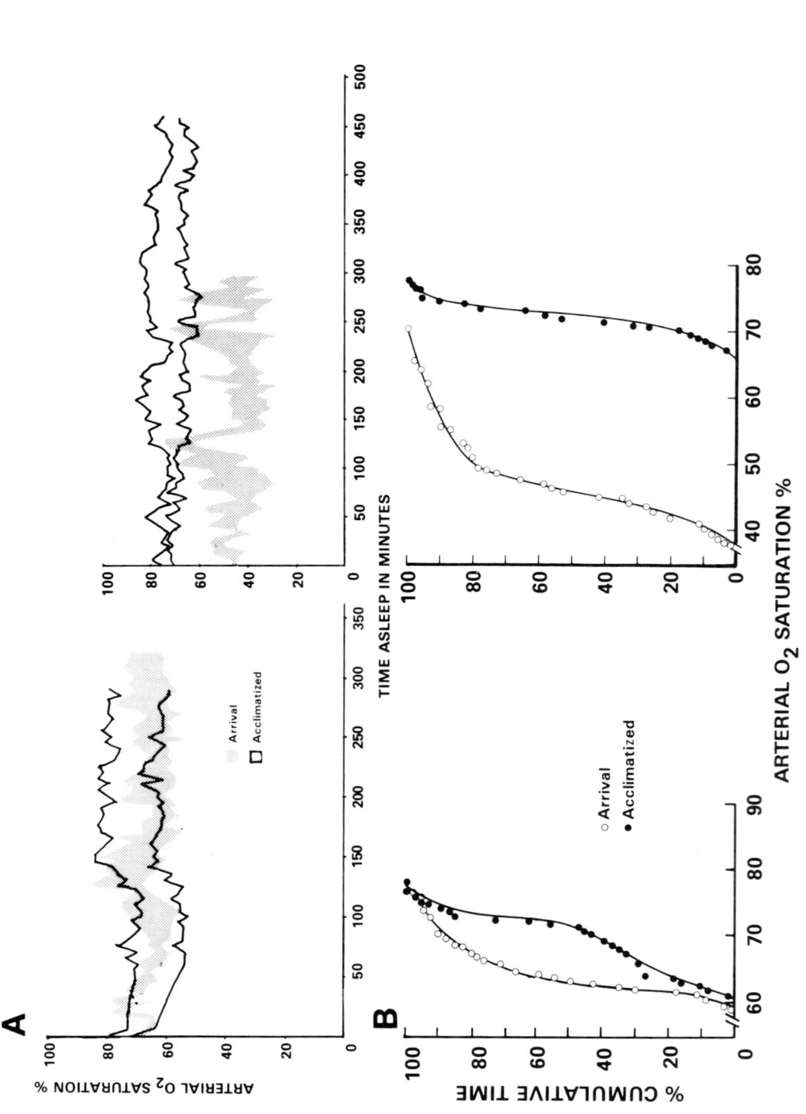

FIG. 2. Sleep saturation at 5,360 m. Subjects in group II (2 males) who were airlifted to 5,360 m and studied 1 and 3 nights after arrival and 8 and 11 days later, respectively. *A* and *B* are similar to those in Fig. 1. [From Powles and Sutton (6).]

TABLE 1. *Arterial oxygen saturation for 50% of night*

Subject	Sex	Age	First Study		Second Study	
			%	Days at 5,360 m	%	Days until 2nd study at 5,360 m
Group I						
1	Female	22	68	5	78	8
2	Female	25	70	2	67	5
3	Male	32	68	3	75	10
Group II						
4	Male	29	64	1	72	11
5	Male	36	46	3	73	8

Values obtained with method of Slutsky and Strohl (7).

a one-tailed t test with an α of 0.05 and a β of 80%, we would require 25 students in order to show that an effect of the order observed in this study was statistically significant.

DISCUSSION

As this study demonstrates, the marked hypoxemia that may occur during sleep in normal subjects at high altitude improves with acclimatization. With one exception (subject *2*) arterial oxygen saturation was better after a period of acclimatization; the differences in her arterial oxygen saturation were less than 5%. In contrast subject *5* had a dramatic increase in arterial oxygen saturation of more than 25%. This subject and subject *4* achieved the same degree of arterial oxygen saturation after they had acclimatized for a further 8 and 11 days, respectively. By this stage the mild symptoms of AMS had disappeared. It is tempting to relate the severity of AMS to the degree of sleep hypoxemia. As before (5), however, we were unable to do this in our study. Furthermore in some subjects we have observed that severe hypoxemia persists even after 6 wk at this altitude without persistence of the symptoms of AMS.

The logistics of the present study kept us from obtaining information in unacclimatized individuals immediately after arrival at altitude. Previously we demonstrated in a hypobaric chamber that arterial oxygen saturation was lower during the first night at a simulated altitude of 4,200 m (10) than the value observed in partially acclimatized individuals at 5,360 m [53.3 ± 5.85 vs. 65.0 ± 11.58 (mean ± SD); $P < 0.05$]. In the only other study comparable with ours, Berssenbrugge and co-workers (1) demonstrated improvement in sleep arterial oxygen saturation and diminution of periodic breathing within the same individuals on consecutive nights.

We previously reported no difference in sleep arterial oxygen saturation in two groups of subjects studied at 5,360 m (10) and have indicated that this may have been due to the large variations between individuals in sleep arterial oxygen saturation. Apparently some individuals may continue to have very low arterial oxygen saturation during sleep, yet be asymptomatic and have no greater risk of retinal hemorrhage (4).

CONCLUSION

Sleep hypoxemia is universal at high altitudes. In accordance with the findings of Berssenbrugge et al. (1), sleep hypoxemia in our study did improve with prolonged stay at altitude in 4 or 5 subjects, and we suggest that this may reflect the ventilatory acclimatization to altitude. The analyses show the clear advantage of using the Slutsky and Strohl method to quantify episodic hypoxemia, especially when studying the effect of an intervention like acclimatization.

We wish to thank Mary Basalygo and Nancy Keech for their excellent technical assistance.

REFERENCES

1. BERSSENBRUGGE, A., J. DEMPSEY, J. SKATRUD, AND C. IBER. The effect of chronic hypoxia on breathing pattern in wakefulness and sleep (abstr.). *Physiologist* 23: 138, 1980.
2. DEJOURS, P., R. H. KELLOGG, AND N. PACE. Regulation of respiration and heart rate response in exercise during altitude acclimatization. *J. Appl. Physiol.* 18: 110–120, 1963.
3. McEVOY, J. D. S., AND N. L. JONES. Arterialised capillary blood gases in exercise studies. *Med. Sci. Sports* 7: 312–315, 1975.
4. McFADDEN, D. M., C. S. HOUSTON, J. R. SUTTON, A. C. P. POWLES, G. W. GRAY, AND R. S. ROBERTS. High altitude retinopathy. *J. Am. Med. Assoc.* 245: 581–586, 1981.
5. POWLES, A. C. P., G. COATES, G. GRAY, K. STROHL, AND J. R. SUTTON. Sea level ventilatory responsiveness, acute mountain sickness and sleep hypoxemia on acute exposure to altitude (abstr.). In: *Hypoxia: Man at Altitude*, edited by J. R. Sutton, N. L. Jones, and C. S. Houston. New York: Thieme-Stratton, 1982, p. 196.
6. POWLES, A. C. P., AND J. R. SUTTON. Sleep at altitude. In: *Seminars in Respiratory Medicine. Man at Altitude.* New York: Thieme-Stratton, 1983, vol. 5, p. 175–180.
7. SAUNDERS, N. A., A. C. P. POWLES, AND A. S. REBUCK. Ear oximetry: accuracy and practicability in the assessment of arterial oxygenation. *Am. Rev. Respir. Dis.* 113: 745–769, 1976.
8. SLUTSKY, A. S., AND K. P. STROHL. Quantification of oxygen saturation during episodic hypoxemia. *Am. Rev. Respir. Dis.* 121: 893–895, 1980.
9. SUTTON, J. R., A. C. BRYAN, G. W. GRAY, E. S. HORTON, A. S. REBUCK, W. WOODLEY, I. D. RENNIE, AND C. S. HOUSTON. Pulmonary gas exchange in acute mountain sickness. *Aviat. Space Environ. Med.* 47: 1032–1037, 1976.
10. SUTTON, J. R., G. W. GRAY, C. S. HOUSTON, AND A. C. P. POWLES. Effects of duration at altitude and acetazolamide on ventilation and oxygenation during sleep. *Sleep* 3: 455–464, 1980.
11. SUTTON, J. R., C. S. HOUSTON, A. L. MANSELL, M. McFADDEN, P. HACKETT, AND A. C. P. POWLES. Effect of acetazolamide on hypoxemia during sleep at high altitude. *N. Engl. J. Med.* 301: 1329–1331, 1979.
12. SUTTON, J. R., A. C. P. POWLES, G. W. GRAY, J. KANE, A. MANSELL, M. McFADDEN, M. ROBERTSON, P. RONDI, AND C. S. HOUSTON. Arterial hypoxemia during maximum exercise at altitude (abstr.). *Clin. Res.* 25: 673A, 1977.
13. WEIL, J. V., H. KRYGER, AND C. H. SCOGGIN. Sleep and breathing at high altitude. In: *Sleep Apnea Syndromes*, edited by C. Guilleminault and W. C. Dement. New York: Liss, 1978, p. 119–123.

13

Respiratory Control in Andean and Himalayan High-Altitude Natives

SUKHAMAY LAHIRI

Department of Physiology, Institute for Environmental Medicine, University of Pennsylvania School of Medicine, Philadelphia, Pennsylvania

STUDIES OF THE HIGH-ALTITUDE NATIVES in the South American Andes focused attention for the first time on aspects of respiratory adaptation that were different from those of sojourners at the same high altitude (4, 9, 12, 28; see also 7, 14, 15). Previously it was thought that the level of adaptation in the fully acclimatized sojourners would be the same as in the native high-altitude residents (1; see also 9). To understand high-altitude adaptation, researchers at the Andean Institute of Biology in Peru compared sea-level natives and high-altitude natives in their own respective environments. Recently respiratory control has gained particular attention because of the striking observation that the adult natives of high altitude ventilate less at a given resting metabolic rate so that their partial pressure of carbon dioxide in alveolar gas ($P_{A_{CO_2}}$) is higher and partial pressure of oxygen in alveolar gas ($P_{A_{O_2}}$) lower (4, 13, 14). It was also established that hyperpnea of exercise was less in the high-altitude natives than in the sojourners (14, 15).

The observations on high-altitude natives in the Himalayas are sparse. Ramaswamy (32), working with Sherpa high-altitude natives at 3,500 m in Kashmir, found that their exercise hyperpnea was significantly less than the newcomers'. Pugh et al. (30) made similar observations on a Sherpa high-altitude native acclimatizing at 5,880 m. A later expedition focused on the mechanism of this phenomenon and control of breathing (17). This and several subsequent studies showed that the Sherpa high-altitude residents manifested blunted ventilatory responses to hypoxia (3, 17–19; see also chapt. 7), with the exception of one report (10). These ventilatory response characteristics resemble those of the natives of high plateaus of the Andes (4, 9, 12, 35; see also 7, 14, 15). This article reviews more recent observations and examines some of those made at 5,400 m during the 1981 American Medical Research Expedition to Everest.

RESTING ARTERIAL AND CENTRAL STIMULI

Alveolar Oxygen and Carbon Dioxide Partial Pressure

Rahn and Otis (31) used a $P_{A_{O_2}}$-$P_{A_{CO_2}}$ diagram to compare alveolar ventilation caused by acute and chronic hypoxia in the sea-level native. This elegant method has many interesting features. For example, it can indicate how the respiratory quotient influences acclimatization to hypoxia (see Fig. 1). An increase in the respiratory quotient at the same inspired P_{O_2} ($P_{I_{O_2}}$) would raise $P_{A_{O_2}}$ and $P_{A_{CO_2}}$. The original curves of Rahn and Otis (31) on lowlanders at high altitude have been modified and extended to the summit of Mount Everest according to the recent data of West et al. (42). The alveolar gas pressures in the high-altitude natives at altitude, taken from the literature

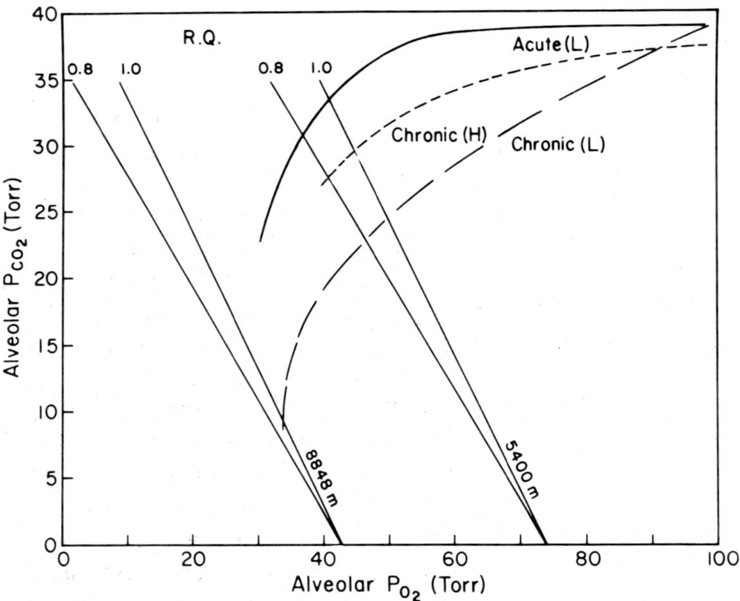

FIG. 1. Alveolar gas composition (P, partial pressure) in sea-level native (L) acutely and chronically exposed to various altitudes compared to that in high-altitude native (H) chronically exposed to similar altitudes. Isoaltitude lines for respiratory quotient (RQ) are 0.8 and 1.0 at 5,400 m (Everest Base Camp) and 8,848 m (Everest summit).

(4, 12, 13, 16, 25; see also 14, 15) and unpublished observations (S. Lahiri), are in the intermediate range between nonacclimatized and acclimatized lowlanders. The Sherpas are also represented by the intermediate curve. Most of the data were collected at various altitudes from the volunteers who did not belong to the selective group of high-altitude Sherpas. However, the natives of the Andes who apparently suffered from chronic mountain sickness provided a different picture (35); they showed practically the same PA_{CO_2} as at sea level despite a much lower PA_{O_2}. These extreme alveolar gas pressures have not been recorded in the Sherpa population studied so far. On the other hand, high-altitude natives show relative hyperventilation at low altitude or on raising PI_{O_2} (12, 13, 15).

Acid Base of Arterial Blood and Cerebrospinal Fluid

Because the stimulating effect of low PO_2 on arterial chemoreceptors is most significantly modified by arterial PCO_2 (Pa_{CO_2}) and [H^+] (e.g., 20), it is critical to measure the arterial stimulus level although most desirable to know the chemosensory input and its relationship with the ventilatory output. The level of stimulus in the milieu of central chemoreceptors is also important.

The Pa_{CO_2} in the Sherpas and high-altitude natives of the Andes is higher than in the sojourners (7, 14, 15). Their Pa_{O_2} is not correspondingly low because of two factors: their respiratory quotient is higher, and presumably their pulmonary diffusing capacity is greater (7, 12, 14, 15). The blood and cerebrospinal fluid are slightly more acid (7, 14, 15). This acidity corresponds to higher PCO_2 in both blood and cerebrospinal fluid due to a hypoventilation relative to metabolic rate.

Scores of acid-base data (7) suggest that full compensation of respiratory alkalosis seldom occurs in the sojourners at high altitude. In the high-altitude native this full compensation occurs perhaps because the respiratory alkalosis is less severe.

RESTING VENTILATORY RESPONSES

Steady-State Response

The steady-state ventilatory responses to carbon dioxide at various levels of PA_{CO_2} for a sojourner and a Sherpa acclimatized at 4,880 m are shown in Figure 2. With the decreasing PA_{O_2} the response curve steepened for the sojourner, whereas it made little difference in the Sherpa whether PA_{O_2} was 200 mmHg or 30 mmHg. Thus the hypoxic response as defined by the slopes of the response curves was severely attenuated in the Sherpa relative to the sojourner. The response to carbon dioxide during hyperoxia was only marginally smaller in the Sherpa. Similarly attenuated steady-state ventilatory responses to hypoxia have been reported for adult male natives of the Peruvian Andes (35; see also 7, 14, 15).

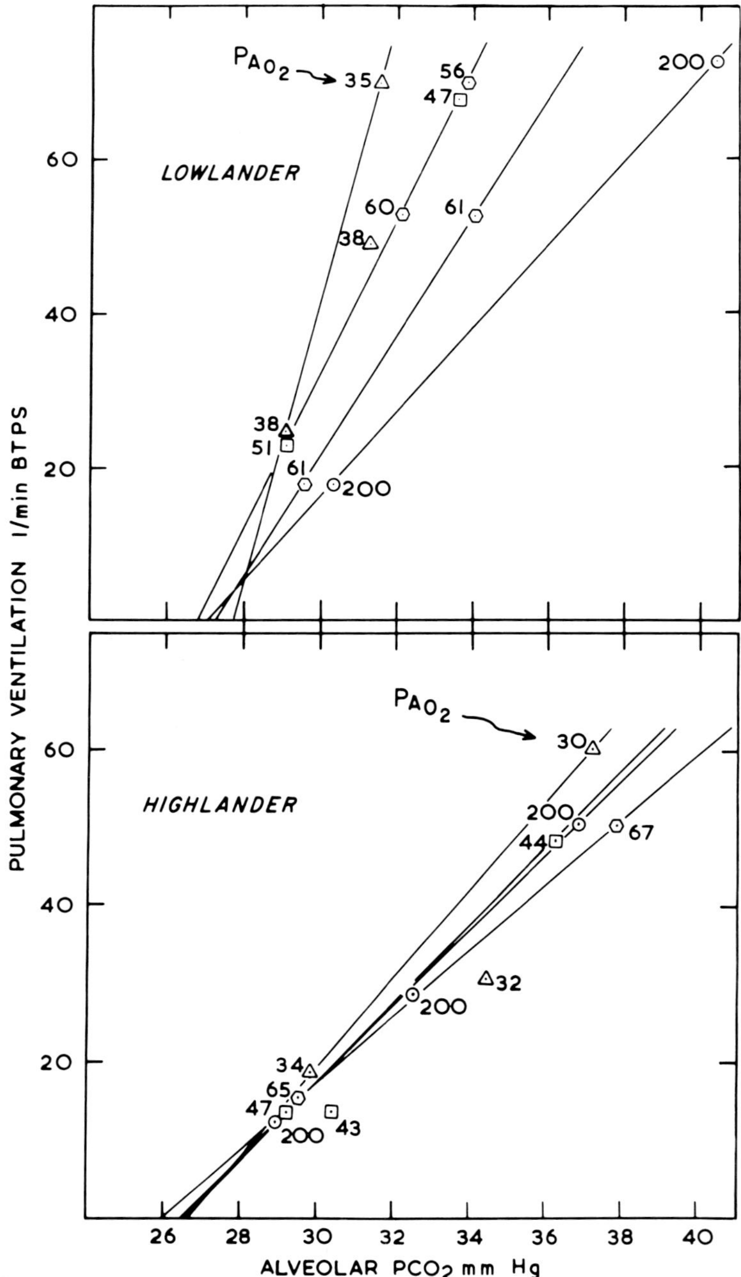

FIG. 2. Steady-state ventilatory responses to various levels of alveolar P_{CO_2} and P_{O_2} ($P_{A_{CO_2}}$ and $P_{A_{O_2}}$) at 4,880 m. Effect of low $P_{A_{O_2}}$ is significantly less in Sherpa highlander (*lower panel*) than in acclimatized lowlander (*upper panel*). [From Milledge and Lahiri (26).]

There are at least two focal points in the chemoreflex pathway at which the attenuation may occur: during the chemosensory responses of aortic and carotid bodies and in the central medullary areas. The excitatory responses to hypoxia and the hypoxic-hypercapnic interaction occur at the level of the peripheral chemoreceptors (8, 24). Accordingly the attenuation can occur at this source. Hypoxia also increases medullary blood flow, which could diminish the local stimulus level and hence the response (7). A depressant effect of hypoxia on the central nervous system also could be a factor (14, 15). If these latter mechanisms are relevant they are probably more so in the adult highlanders than in the sojourners, but the highlanders tolerate extreme hypoxia better (40).

Transient-State Response

Measuring ventilatory responses to brief changes in PA_{O_2} and Pa_{O_2} is advantageous because the response occurs before the PO_2 and PCO_2 changes influence the central nervous system. However, the arterial chemoreceptor activity during the transient change in Pa_{O_2} is probably not the same as that in the steady state, and the ventilatory response may not fully correspond to the chemoreceptor activity (see 8).

The ventilatory responses to lowering the percent of arterial oxygen saturation (Sa_{O_2} %) with a few breaths of inhaled nitrogen are compared between a lowlander and a Sherpa in Figures 3 and 4. Clearly the two responses differ: the lowlander responded vigorously, whereas the Sherpa showed only a small increase in ventilation. The responses of an average lowlander and a Sherpa are compared in Figure 5. Transient oxygen breathing and an increased Sa_{O_2} % made little difference to the Sherpa's ventilation.

The transient responses of the Bolivian and Peruvian high-altitude natives have also been measured (7, 14). These data are in agreement, showing that their responses to hypoxia are attenuated.

Non-Steady-State Response by Carbon Dioxide Rebreathing

This method is an adaptation of that originally described by Read (34). The seated subject rebreathed carbon dioxide from a small bag at 5,400 m. Initially the bag contained 7% carbon dioxide in air or oxygen. The end-tidal PO_2 (PET_{O_2}) was held constant by replenishing oxygen in the bag while the carbon dioxide level increased gradually. The responses of a Sherpa and a Caucasian lowlander who climbed Mount Everest on October 21, 1981 are compared in Figure 6. The following differences are striking: *1*) the Sherpa showed a relatively small ventilatory difference between the hypoxic and hyperoxic carbon dioxide responses, and *2*) the slopes of the hypoxic and hyperoxic carbon dioxide response curves for the Sherpa were 2.4 and 1.5 liters$\cdot$min$^{-1}\cdot$mmHg^{-1} PET_{CO_2}, respectively, in contrast to 4.3 and 2.2 liters$\cdot$min$^{-1}\cdot$mmHg^{-1} PET_{CO_2} in the lowlander. Definitely both the level of ventilation at any given PA_{O_2} and the ventilatory response for any given PA_{O_2} change require analysis. The differences in the levels of ventilation and the slopes of the response curves at two levels of PA_{O_2} were smaller in the Sherpa than in the lowlander, indicating a lower hypoxic ventilatory drive. The Sherpa's

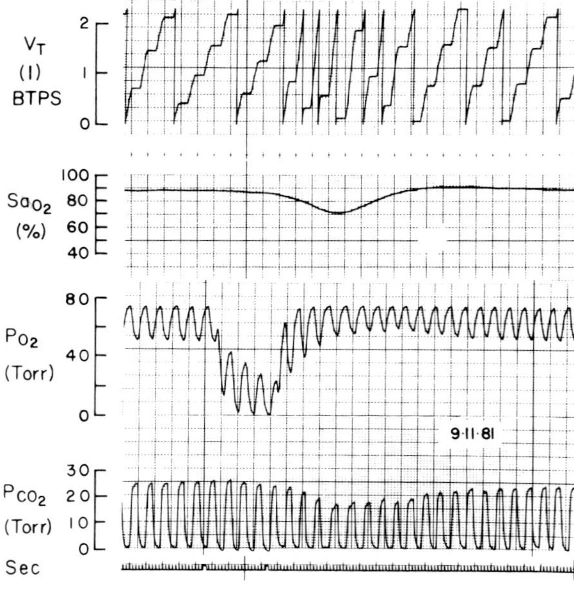

FIG. 3. Ventilatory response to transient acute hypoxia in a sojourner at 5,400 m. VT, ventilation; Sa_{O_2}, arterial O_2 saturation; P_{O_2}, mouthpiece P_{O_2}; P_{CO_2}, mouthpiece P_{CO_2}.

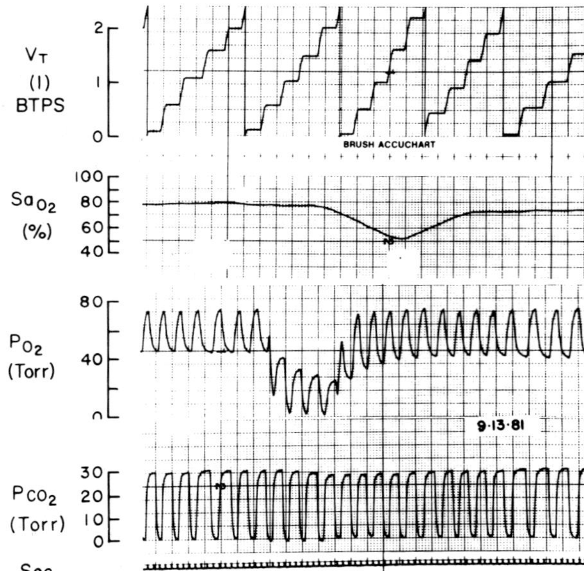

FIG. 4. Ventilatory response to transient acute hypoxia in a Sherpa at 5,400 m.

ventilatory response to carbon dioxide in hyperoxia is also low. These characteristics of chemical drives are not consistent with his high ventilation at any $P_{A_{O_2}}$. This subject might have hyperventilated when using the breathing valve and mouthpiece, raising his absolute ventilation artificially.

The distribution of ventilatory sensitivity to hypoxia among adult high-altitude natives in the Himalayas and Andes has been compared to that of sea-level natives. The units of these indices measured by various methods are not the same, however. For comparison these units are all expressed as percents

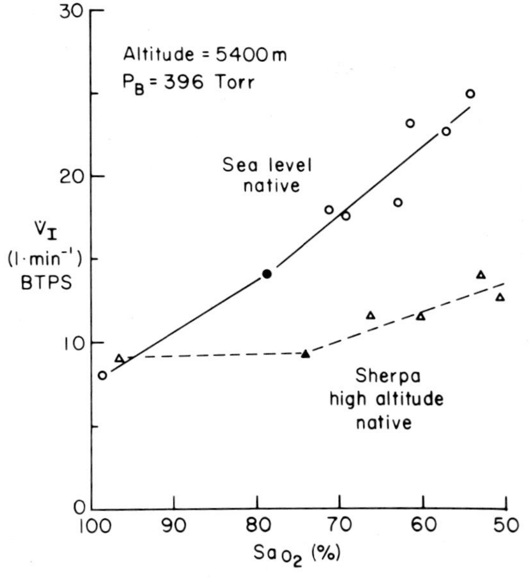

FIG. 5. Comparison of ventilatory responses ($\dot{V}_I$) to various levels of Sa_{O_2} % between a Sherpa and a sojourner at 5,400 m. These are transient responses to acute hypoxia and hyperoxia.

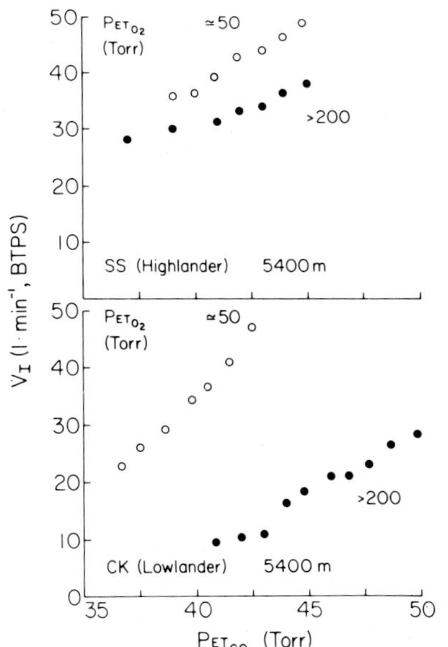

FIG. 6. Comparison of ventilatory responses to iso-oxic CO_2 rebreathing (PET, end-tidal partial pressure) between a Sherpa (*upper panel*) and a sojourner (*lower panel*) at 5,400 m.

of the maximal response reported in each study (3, 4, 12, 13, 16, 18, 19, 25, 26, 30, 34, 35; see also chapt. 7). Each study was performed at high altitude and included both natives to high altitude and sea-level natives acclimatized to high altitude. The frequency distribution of all these values (%) is shown in Figure 7. New data from the 1981 Everest Expedition are also included. Both Andean and Himalayan high-altitude natives show attenuated responses compared to the lowlanders acclimatized to the same altitudes. No high-altitude native had a hypoxic ventilatory sensitivity above 80% of the lowlander maximum.

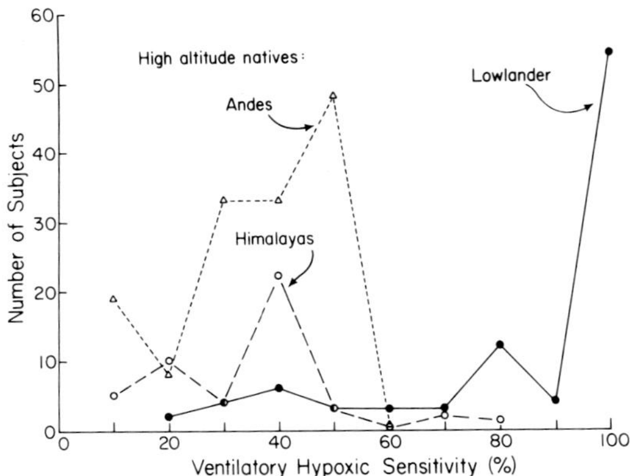

FIG. 7. Frequency distribution of hypoxic drive for ventilation at high altitude in acclimatized sea-level natives, Sherpas in the Himalayas, and Indians in the Andes.

Polycythemia and Hypoxic Chemosensitivity

High-altitude polycythemia in humans, a well-known phenomenon, originates from the stimulation of erythropoiesis by tissue hypoxia. Severinghaus et al. (35) showed that the large difference in the ventilatory sensitivity to hypoxia between sojourners and normal natives to 4,330 m in the Andes was associated with a small difference in hematocrit, whereas the highlanders with only small differences in hypoxic sensitivity showed a large spread of hematocrit. Lahiri and Milledge (17) found that, although Sherpa highlanders had lower hemoglobin than the lowlanders after acclimatization at 4,800 m, their ventilatory sensitivity to hypoxia was also lower. The Sherpa residents at altitudes between 3,000 m and 4,000 m had higher blood hemoglobin contents than residents at lower altitudes in Nepal and usually lower than the acclimatized lowlanders at the same altitude (S. Lahiri, unpublished observations). Also the natives of Andean high altitude who migrated to sea level lost polycythemia within a few weeks but did not regain normal ventilatory sensitivity to hypoxia within 10 mo of their residence at sea level (9, 12, 14). The hypoxic sensitivity of sea-level natives who migrated to high altitude had not changed when they developed polycythemia (7, 14, 15). Also the young natives of high altitude who show normal hypoxic chemosensitivity comparable to the sea-level children have a higher hematocrit (7, 14, 15). Therefore the attenuated ventilatory sensitivity to hypoxia is not due to polycythemia. On the other hand, some subjects who show very low hypoxic stimulation of ventilation have excessive polycythemia at a given high altitude: low Pa_{O_2}, high P_{CO_2}, and low pH (35). The consensus is that the polycythemia is a result rather than a cause of the insensitivity. The oxygen-sensitive erythropoietic tissues (and perhaps other tissues) are responsive to hypoxia in these subjects even though their chemoreflex is blunted. Also some animals acclimatize well at high altitude with little or no significant erythropoietic response (14).

Chemosensory responses from carotid bodies are not known to be sensitive to limited changes in hematocrit, whereas those from aortic chemoreceptors are (2, 21; see also 8, 24). However, ventilatory chemoreflex responses to hypoxia depend more on carotid body chemosensory responses than on those from aortic body chemosensors (8).

Diseases of Ventilatory Control System

Each component in the control system is a potential point for disorder. Our interest in chronic hypoxia is with the chemoreflex pathway. Abnormal growth of carotid body (chemodectoma) is a known disease, occurring most frequently at high altitude (3). However, we know nothing of its functional consequences. On the other hand, Andean high-altitude natives who manifest chronic mountain sickness with loss of hypoxic ventilatory drive (9, 12, 28, 35) seldom display chemodectoma. By exclusion one may suggest that chronic mountain sickness is a pathology of the chronically hypoxic central nervous system. The disease, however, is not found at all high altitudes but is most prevalent in the industrial and mining areas of North America and South America (3,100 m–4,540 m).

The ventilatory sensitivity to carbon dioxide, although also diminished in chronic mountain sickness, is adequate for survival. A loss of carbon dioxide sensitivity in these patients along with hypoxic ventilatory sensitivity would be fatal. This is perhaps a reason why subjects with a loss of carbon dioxide sensitivity have not been reported.

Hyperventilation on Oxygen Breathing

The sojourners at high altitude decrease their ventilation by breathing 100% oxygen, and it remains diminished in steady-state hyperoxia. The Andean high-altitude natives, on the other hand, hyperventilate with steady-state hyperoxia (12, 16, 35). The Sherpas also show relative hyperventilation during steady-state hyperoxia (10, 17, 26). Incidentally this hyperventilation on oxygen breathing gives an edge to the Sherpas who climb to extreme altitude breathing supplemental oxygen.

When moving to lower altitudes the Sherpas and Peruvian high-altitude natives show relative hyperventilation (7; see also 14, 15). Consequently the spread of ventilation due to the same changes in altitude is smaller in the high-altitude natives. The mechanism of this hyperoxic hyperventilation in the high-altitude natives is not clear. However, it may be related to the blunted chemoreflex drive because cats after carotid sinus nerve section lack a ventilatory response to hypoxia and show a hyperventilatory response to hyperoxia (27). Hyperoxic hyperventilation in the high-altitude natives has also been interpreted as reversal of hypoxic depression (16, 35).

Time-Dependent Change

The attenuated ventilatory sensitivity to hypoxia in adult natives of high altitude could be genetic. The question has been investigated in natives of high altitudes in the Peruvian Andes and the Rockies. Weil et al. (41) found that the young children at 3,100 m in the Rockies manifested normal sensitivity, whereas the adults showed low sensitivity; they concluded that the latter was acquired with time. My colleagues and I have shown that the newborns of a few days age at 3,850 m in the Andes manifested little ventilatory response to 100% oxygen breathing, whereas ventilation in older infants and young children was strikingly decreased (14). Similar observations were made on the newborns and infants of the same native population at 800 m. Consistent with observations at sea level (14, 15), the conclusion was that newborns, regardless of the altitude of their birth, exhibit a low ventilatory response to hypoxia. Therefore the full ventilatory response to hypoxia in the normal subjects appears to develop after birth.

A decline in this response occurs at a later stage in life. This phenomenon in the natives of the Peruvian Andes is shown in Figure 8. The peak response, which had already occurred by 6 yr of age, was followed by a decline. A decline in the sensitivity also occurs in the sea-level natives much later in life and is considered to be an effect of aging. It appears though that this aging effect occurs early in adult life at high altitude, although that need not be true.

Studies show that the adult natives of high altitude who migrate to sea level do not regain their sensitivity in a year or two. This could lead to the conclusion that the phenomenon is irreversible (14). However, population studies provide evidence that with a longer residence at sea level the phenomenon is reversed (7, 14). A corollary to this observation is that the aging effect presumably does not cause attenuation of the response at high altitude; nor is it likely that the low responders are those that stay at high altitude past adolescence.

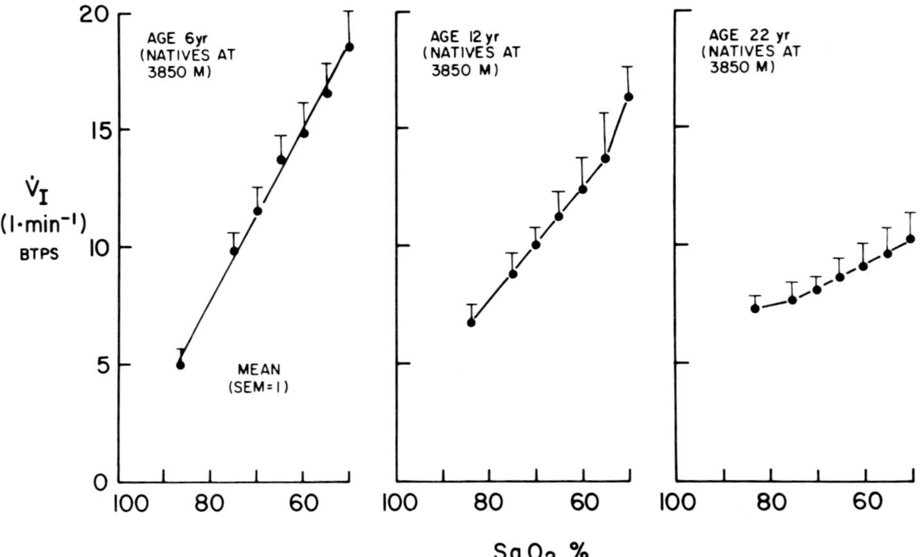

FIG. 8. Time-dependent decline of ventilatory response to acute hypoxia in Andean high-altitude natives at 3,850 m (Puno, Peru).

Controversy

With the premise that to perform well at high altitude an acclimatized individual must hyperventilate adequately, Hackett et al. (10) investigated Sherpa high-altitude residents in the Himalayas; they concluded that the Sherpas do manifest a high resting ventilation. They also reported ventilatory responses to hypoxia less attenuated than previously reported by Ramaswamy (32), Pugh et al. (30), Ceretelli (3), and Lahiri and colleagues (17–19, 28). Because of the controversy it is appropriate to briefly examine the observations of Hackett et al. (10). Their Sherpa subjects at 1,377 m hyperventilated, as found by others at sea level (7, 14, 15); at 4,243 m they hyperventilated when breathing 100% oxygen, as did the Sherpas in the studies of Lahiri and colleagues. High-altitude natives in the Andes also hyperventilate on breathing 100% oxygen (32, see 13, 14). Thus the span of ventilatory responses to the range of $P_{I_{O_2}}$ from hypoxia to hyperoxia was small in the Sherpa and Andean high-altitude native.

Hackett et al. (10) did find that the hypoxic ventilatory sensitivity in the Sherpas declined with increasing age and altitude exposure. Clearly Sherpas who were older than 20 yr and had resided at or above 3,800 m developed a blunted hypoxic response. That they observed no periodic breathing with apnea during sleep in Sherpas further strengthens this hypothesis. According to our interpretation this lack of periodic breathing is due to their blunted hypoxic drive to ventilation (20). The major point of Hackett et al. (10) is that all their Sherpa subjects hyperventilated at any altitude studied. However, they also reported that this resting hyperventilation was not correlated with their hypoxic ventilatory sensitivity. Because the measurement of resting ventilation is often difficult (particularly on untrained Sherpas), the subjects may hyperventilate while being equipped with the respiratory apparatus. Accordingly the controversy is more apparent than real. Further studies on Sherpas are desirable, but their migration and frequent trips to lower altitudes may alter their adaptation characteristics with time.

HYPERPNEA OF HYPOXIC EXERCISE

Exercise at sea level increases ventilation in proportion to the metabolic rate so that $P_{A_{CO_2}}$ and Pa_{CO_2} remain practically unchanged. Accordingly a given hypoxic drive that increases resting ventilation and lowers $P_{A_{CO_2}}$ would increase ventilation during exercise by a factor by which carbon dioxide production is increased over the resting metabolic rate. The site and mechanism of this multiplicative effect are not known primarily because the mechanism of exercise hyperpnea is not known. However, the mean activity of carotid body chemoreceptors is not increased during exercise (8) although the phasic activity may provide part of the explanation (5, 6). In any case the augmented ventilatory effect of hypoxia during exercise depends on the peripheral chemosensory input. Thus ventilation in hypoxic exercise provides a test for the peripheral chemosensory drive. According to the test, Sherpas and other high-altitude natives show blunted hypoxic drive (18, 19, 30, 32; see also 7, 14, 15). Consistent with this, raised $P_{I_{O_2}}$ lowers ventilation more in the sojourners than in the highlanders at high altitude.

The time course of ventilatory responses to hypoxic exercise at 5,400 m showed delayed and slow increases in highlanders compared with those in the sojourners. Also at high altitude raised $P_{I_{O_2}}$ made a small difference in the responses in natives but greatly delayed and diminished those in sojourners. These data confirm the role of an attenuated peripheral chemoreflex in the control of exercise hyperpnea in the Sherpas at high altitude.

VENTILATION DURING SLEEP

Sleep or somnolence in the sojourners at high altitude leads to periodic breathing with apnea. This characteristic depends on a high ventilatory sensitivity to hypoxia (see chapt. 7). The Sherpas seldom showed periodic breathing with apnea during sleep at 5,400 m. Also a raised $P_{I_{O_2}}$ in the sojourners stopped their periodic breathing with apnea. This common response suggests that an attenuated peripheral chemosensory drive is responsible for the lack of sleep apnea with periodic breathing in the Sherpas. The effect of sleep on ventilation among the Andean high-altitude natives has not been reported, nor do we know the breathing pattern in the young high-altitude natives who show normal hypoxic responses.

MECHANISMS

Adaptation is apparent in the changes of response threshold and sensitivity to a given chronic stimulus. Tenney and Ou (39) believe that effects are blunted by an increase in the stimulus threshold.

Attenuated stimulus threshold and sensitivity of the chemoreflexes in breathing may originate in one or more sites in the reflex pathway. The peripheral chemosensors are the ideal candidate for this role, but the phenomenon may occur even with a normal chemosensory response. Aortic body chemosensors have a lower hypoxic threshold and sensitivity than the carotid body chemoreceptors presumably because of their relative insensitivity to carbon dioxide and hydrogen ion (20). The mechanism of the latter insensitivity is unknown, but possibly the carotid body chemosensors in the high-altitude natives respond like the aortic body chemosensors. A second mechanism relies on endogenous dopamine. Dopamine attenuates carotid and aortic body chemosensory responses to hypoxia (22, 29, 36), it is present in the carotid and aortic bodies (8), its concentration and total amount are increased severalfold during chronic hypoxia (8, 24), and it is released by hypoxia (8). A third mechanism could be provided by raising tissue P_{O_2} at the receptor site relative to the arterial blood. The favorable factors are increased blood and vascular volumes in the carotid body and polycythemia. Despite these very plausible mechanisms the carotid body chemosensory responses have not been found to be attenuated. Lahiri et al. (23) found that carotid body chemosensors from the intact sinus nerve did not show an attenuated response to hypoxia in cats exposed to chronic hypoxia ($P_{I_{O_2}}$ = 70 mmHg) over several weeks, although some of these cats did show attenuated ventilatory responses to hypoxia (P. Barnard and S. Lahiri, unpublished observations), as reported by Tenney and Ou (39). Thus the phenomenon of blunting seems to reside in the chemoreflex pathway beyond the peripheral chemoreceptors. This conclusion is consistent with the work of Tenney and his colleagues (39; see also 7, 14).

There is a distinction between the chronically hypoxic models for the human and the cat. Adult human sojourners exposed to 5,880 m or higher altitudes during a climb are not known to develop a blunted ventilatory response to hypoxia in a few weeks. Young human high-altitude natives do not show blunted ventilatory drives either. However, the structural and biochemical changes in the carotid and aortic bodies due to chronic hypoxia will have taken place at high altitude, indicating that they do not cause the blunting.

Comparing the ventilatory responses to hypoxia and hypercapnia shows that the hypoxic response is significantly more attenuated than the hypercapnic response (26, 35). This suggests that the hypoxic chemoreflex pathway is specifically involved. The two chemosensory inputs from separate pathways that converge onto the final common neurons in the brain stem may be the morphological basis for the separation of effects. St. John (38) reported that all respiratory neurons are not stimulated by both peripheral and central chemosensory input. Also haloperidol (1 mg/kg iv), a dopamine antagonist that crosses the blood-brain barrier, attenuates the ventilatory response to steady-state hypoxia without significantly altering that to hypercapnia in anesthetized cat (37). This blunting of ventilatory response to carotid chemosensory input is illustrated in Figure 9. The *upper panel* shows the relationship at several levels of Pa_{O_2} at a constant Pa_{CO_2} and the *lower panel* at various levels of Pa_{CO_2} in hyperoxia. Clearly the ventilatory response to the increases in carotid chemosensory input caused by hypoxia is attenuated, whereas that caused by hypercapnia is not. The latter is attributed to the centrally mediated chemosensory effect of carbon dioxide and hydrogen ion. The peripheral chemosensory input at a given Pa_{O_2} is increased after haloperidol administration (22, 36). Ventilation remains depressed despite this increased peripheral chemoreceptor response. This blockade probably occurred in the chemoreflex pathway to respiratory neurons, which are also not common to the central carbon dioxide and hydrogen ion chemosensory input. The hypothesis is that the dopamine receptors in this brain stem pathway mediate the excitatory input from the peripheral chemoreceptors and hence may be a site for the blunting of chemoreflex. This possible mechanism and that proposed by Tenney are not mutually exclusive because the distal cortical inhibition may depend on the proximal site in the brain stem.

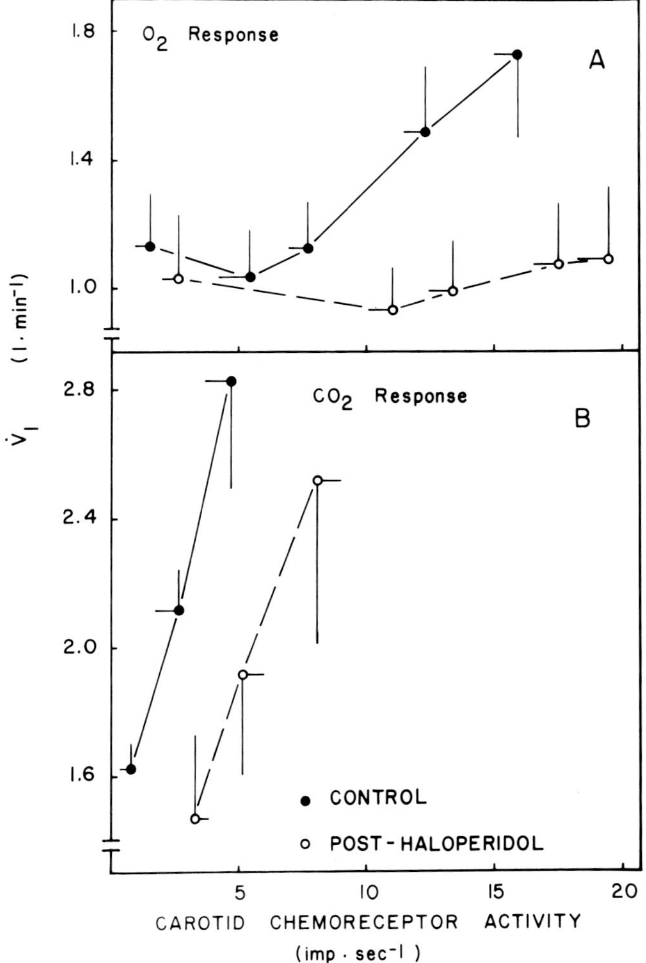

FIG. 9. Effect of haloperidol on ventilatory responses to hypoxia (*A*) and hypercapnia (*B*) in anesthetized cat. Ventilation is plotted against carotid chemosensory input. Blunting of steady-state ventilatory response to hypoxia despite increases in chemosensory discharge rates from carotid body after haloperidol is clear. Effect of hypercapnia (slope) remained substantially intact.

STRATEGY FOR ADAPTATION

Ventilation is a part of the transport system for oxygen from the source (ambient air) to the sink (the cells) where it is consumed. With a decreased Po_2 in the air at high altitude, the pressure gradient driving oxygen to the cell is diminished. As a result physiological responses and adaptations can occur both in the oxygen transport system and in the cellular metabolic machinery. Numerous examples in nature show structural and functional changes in the same species according to their distinctive environment. These examples point to the principle of conservation, which is a balance between optimization of energy production and minimization of its cost, increasing the efficiency of the system. These adjustments can occur in any cell and organ, between organs, and in the interaction between an organism and its environment.

Parallels can be drawn between living at high altitude and living without oxygen. Hochachka (11) surveyed and analyzed organisms with variations in hypoxia tolerance and illustrated the unifying theme shared by diverse organisms. The multicellular and multienzyme systems have many alternatives. Accordingly alterations in the quality and quantity of fermentable substrates, maintenance of redox balance between mitochondria and cytosol, and changes in the kinetic property and quantity of critical enzymes may occur, extending the organism's anaerobic potential and hypoxic tolerance.

Transport of oxygen by ventilation and circulation incurs energy expenditure without energy production. If the goal of adaptation to hypoxia is to maximize energetics, one way is to minimize the transport expenses for oxygen. The natives of high altitude would presumably do exactly that as their cells and tissues began to use the mechanisms. Changes in the hypoxic ventilatory hyposensitivity and threshold may be a part of this mechanism. This allows a greater ventilatory reserve to achieve a greater work capacity at high altitude (33). This adaptation also maintains stable breathing without wide fluctuations (e.g., during sleep) and maintains homeostasis.

Neither the Andean nor the Himalayan natives to high altitudes habitually climb high mountains for sport, but they perform hard and sustained work at their altitude of residence (12, 28, 30) and tolerate very low $P_{I_{O_2}}$ (40). On the other hand, performance at extreme altitude in field conditions demands skills and training. Compared with the trained climbers the high-altitude natives are at a disadvantage. Yet the Sherpas are an equal match at extreme altitudes. Curiously, however, Andean natives do not share this reputation. It is reasonable to stress that the transport system for oxygen, which expresses the response and bears the heavy burden during the initial phase of altitude exposure, changes with time and plays less of a role where adaptation at the cellular level occurs, reducing the physiological cost. This scheme conforms to the principle of optimization in the integrative control system, a goal of all living processes for sustenance.

The material presented here came from observations on many expeditions to high altitudes, including four Himalayan and two Andean in which I participated. The most recent one was the 1981 American Medical Research Expedition to Everest. I am grateful to John B. West for inviting me to be a member of this expedition. The scientific accomplishments resulted from collaboration of many expedition members. This account, however, expresses my prejudices based partly on my working experience with the high-altitude natives. I am grateful to those who made these observations and experiences possible and to J. Callaghan, K. Hart, B. Pauly, P. Barnard, and A. Mokashi for their assistance.

The work was supported by several grants over many years; the most recent one was National Institutes of Health Grant HL-26533.

REFERENCES

1. BARCROFT, J. B. *The Respiratory Function of the Blood. I. Lessons From High Altitudes.* Cambridge, UK: Cambridge Univ. Press, 1925.
2. BISCOE, T. J., G. W. BRADLEY, AND M. J. PURVES. The relation between carotid body chemoreceptor discharge, carotid sinus pressure and carotid body venous flow.

J. Physiol. London 208: 99–120, 1970.

3. CERETELLI, P. Gas exchange at high altitude. In: *Pulmonary Gas Exchange*, edited by J. B. West. New York: Academic, 1980, vol. II, p. 98–147.
4. CHIODI, H. Respiratory adaptations to chronic high altitude hypoxia. *J. Appl. Physiol.* 10: 81–87, 1957.
5. CROSS, B., D. A. GUZ, P. G. KATONA, M. MACLEAN, K. MURPHY, S. J. G. SEMPLE, AND R. STIDWILL. The pH oscillations in arterial blood during exercise: a potential signal for the ventilatory response in the dog. *J. Physiol. London* 329: 57–73, 1982.
6. CUNNINGHAM, D. J. C. The control system regulating breathing in man. *Q. Rev. Biophys.* 6: 433–483, 1973.
7. DEMPSEY, J. A., AND H. V. FORSTER. Mediation of ventilatory adaptations. *Physiol. Rev.* 62: 262–346, 1982.
8. EYZAGUIRRE, C., R. S. FITZGERALD, S. LAHIRI, AND P. ZAPATA. Arterial chemoreceptors. In: *Handbook of Physiology. Peripheral Circulation and Organ Blood Flow*, edited by J. T. Shepherd and F. M. Abboud. Bethesda, MD: Am. Physiol. Soc., 1983, sect. 2, vol. III, pt. 2, chapt. 16, p. 557–621.
9. FRISANCHO, R. A. *Human Adaptation.* Ann Arbor: Univ. Michigan Press, 1981.
10. HACKETT, P., J. T. REEVES, C. D. REEVES, R. F. GROVER, AND D. RENNIE. Control of breathing in Sherpas at low and high altitude. *J. Appl. Physiol.: Respirat. Environ. Exercise Physiol.* 49: 374–379, 1980.
11. HOCHACHKA, P. W. *Living Without Oxygen.* Cambridge, UK: Harvard Univ. Press, 1980.
12. HURTADO, A. Animals in high altitudes: resident man. In: *Handbook of Physiology. Adaptation to the Environment*, edited by D. B. Dill and E. F. Adolph. Washington, DC: Am. Physiol. Soc., 1964, sect. 4, chapt. 54, p. 843–860.
13. LAHIRI, S. Alveolar gas pressures in man with life-time hypoxia. *Respir. Physiol.* 4: 373–386, 1968.
14. LAHIRI, S. Physiological response and adaptations to high altitude. In: *Environmental Physiology II*, edited by D. Robertshaw. Baltimore, MD: University Park, 1977, vol. 15, p. 217–251. (Int. Rev. Physiol. Ser.)
15. LAHIRI, S., P. BARNARD, AND R. ZHANG. Initiation and control of ventilatory adaptation to chronic hypoxia of high altitude. In: *Control of Respiration*, edited by D. Pallot. London: Helm, 1983, p. 298–325.
16. LAHIRI, S., F. F. KAO, T. VELASQUEZ, C. MARTINEZ, AND W. PEZZIA. Irreversible blunted respiratory sensitivity to hypoxia in high altitude natives. *Respir. Physiol.* 6: 360–374, 1969.
17. LAHIRI, S., AND J. S. MILLEDGE. Sherpa physiology. *Nature London* 207: 610–612.
18. LAHIRI, S., J. S. MILLEDGE, H. P. CHATTOPADHYAY, A. K. BHATTACHARYYA, AND A. K. SINHA. Respiration and heart rate of Sherpa highlanders during exercise. *J. Appl. Physiol.* 23: 545–554, 1967.
19. LAHIRI, S., J. S. MILLEDGE, AND S. C. SØRENSON. Ventilation in man during exercise at high altitude. *J. Appl. Physiol.* 32: 766–769, 1972.
20. LAHIRI, S., A. MOKASHI, E. MULLIGAN, AND T. NISHINO. Comparison of aortic and carotid chemoreceptor responses to hypercapnia and hypoxia. *J. Appl. Physiol.: Respirat. Environ. Exercise Physiol.* 51: 55–61, 1981.
21. LAHIRI, S., E. MULLIGAN, T. NISHINO, A. MOKASHI, AND R. O. DAVIES. Relative responses of aortic body and carotid body chemoreceptors to carboxyhemoglobinemia. *J. Appl. Physiol.: Respirat. Environ. Exercise Physiol.* 50: 580–586, 1981.
22. LAHIRI, S., T. NISHINO, A. MOKASHI, AND E. MULLIGAN. Interaction of dopamine and haloperidol with O_2 and CO_2 chemoreception in carotid body. *J. Appl. Physiol.: Respirat. Environ. Exercise Physiol.* 49: 45–51, 1980.
24. LAHIRI, S., N. J. SMATRESK, AND E. MULLIGAN. Responses of peripheral chemoreceptors to natural stimuli. In: *Physiology of the Peripheral Arterial Chemoreceptors*, edited by H. Acker and R. O'Regan. Amsterdam: Elsevier, 1983, p. 221–256.
25. LEFRANCOIS, R., H. GAUTIER, AND P. PASQUIS. Ventilatory oxygen drive in acute and chronic hypoxia. *Respir. Physiol.* 4: 217–228, 1968.
26. MILLEDGE, J. S., AND S. LAHIRI. Respiratory control in lowlanders and Sherpa highlanders at altitude. *Respir. Physiol.* 2: 310–322, 1967.
27. MILLER, J. D., AND S. M. TENNEY. Hypoxia-induced tachypnea in carotid deafferented cats. *Respir. Physiol.* 23: 31–39, 1975.
28. MONGE, M. C., AND C. C. MONGE. *High Altitude Diseases. Mechanisms and Management.* Springfield, IL: Thomas, 1966.
29. NISHINO, T., AND S. LAHIRI. Effects of dopamine on chemoreflexes in breathing. *J. Appl. Physiol.: Respirat. Environ. Exercise Physiol.* 50: 892–897, 1981.
30. PUGH, L. G. C. E., M. B. GILL, S. LAHIRI, J. S. MILLEDGE, M. P. WARD, AND J. B. WEST. Muscular exercise at great altitudes. *J. Appl. Physiol.* 19: 431–440, 1964.
31. RAHN, H., AND A. B. OTIS. Man's respiratory response during and after acclimatization to high altitude. *Am. J. Physiol.* 157: 445–462, 1949.
32. RAMASWAMY, S. S. Load carriage by infantry soldiers at high altitude. In: *International Symposium on Problems of High Altitude.* New Delhi, India: Armed Forces Medical Services, 1962, p. 74–86.
33. READ, D. J. C. A clinical method for assessing the ventilatory response to carbon dioxide. *Aust. Ann. Med.* 16: 20–32, 1967.
34. SALDANA, M. J., L. E. SALEM, AND R. TRAVEZAN. High altitude hypoxia and chemodectomas. *Hum. Pathol.* 4: 251–263, 1973.
35. SEVERINGHAUS, J. W., C. R. BAINTON, AND A. CARCELÉN. Respiratory insensitivity to hypoxia in chronically hypoxic man. *Respir. Physiol.* 1: 308–334, 1966.
36. SMATRESK, N., AND S. LAHIRI. Aortic body chemoreceptor responses to dopamine, haloperidol, and pargyline. *J. Appl. Physiol.: Respirat. Environ. Exercise Physiol.* 53: 596–602, 1982.
37. SMATRESK, N. J., M. POKORSKI, AND S. LAHIRI. Opposing effects of dopamine receptor blockade on ventilation and carotid chemoreceptor activity. *J. Appl. Physiol.: Respirat. Environ. Exercise Physiol.* 54: 1567–1573, 1983.
38. ST. JOHN, W. M. Respiratory neuron responses to hypercapnia and carotid chemoreceptor stimulation. *J. Appl. Physiol.: Respirat. Environ. Exercise Physiol.* 51: 816–822, 1981.
39. TENNEY, S. M., AND L. C. OU. Hypoxic ventilatory response of cats to high altitude: an interpretation of blunting. *Respir. Physiol.* 30: 185–189, 1977.
40. VELÁSQUEZ, T. Tolerance to acute anoxia in high altitude natives. *J. Appl. Physiol.* 14: 357–362, 1959.
41. WEIL, J. V., E. BYRNE-QUINN, E. INGVAR, G. F. FILLY, AND R. F. GROVER. Acquired attenuation of chemoreceptor function in chronically hypoxic man at high altitude. *J. Clin. Invest.* 50: 186–195, 1971.
42. WEST, J. B., P. H. HACKETT, K. H. MARET, J. S. MILLEDGE, R. M. PETERS, JR., C. J. PIZZO, AND R. M. WINSLOW. Pulmonary gas exchange on the summit of Mount Everest. *J. Appl. Physiol.: Respirat. Environ. Exercise Physiol.* 55: 678–687, 1983.

14

High-Altitude Polycythemia

ROBERT M. WINSLOW

Division of Host Factors, Center for Infectious Diseases, Centers for Disease Control, Public Health Service, United States Department of Health and Human Services, Atlanta, Georgia

THE FRENCH PHYSICIAN Viault traveled by train from Lima to Morococha, a small mining town situated at 4,520 m in the Peruvian Andes, in 1980. There he noted for the first time that the number of red cells increased at high altitude and thereby settled a dispute that had been argued for years about the mechanisms of acclimatization (11). Subsequently C. M. Monge and many of his followers in Peru fully described the polycythemia in high-altitude natives.

A second hematological alteration, an increase in the concentration of the red cell metabolite 2,3-diphosphoglycerate (2,3-DPG), occurs on exposure to hypoxia. This metabolite reduces the affinity of hemoglobin for oxygen, but the in vivo effect is minimal because it is counteracted by a concomitant increase in plasma pH (Bohr effect). Chapter 6 discusses these red cell changes in light of recent data.

Most current descriptions of acclimatization to high altitude cite polycythemia as one of the principal reactions of normal humans. However, many high-altitude natives become severely symptomatic from excessive polycythemia, a condition called chronic mountain sickness (CMS), now commonly known as Monge's disease. These fascinating cases have raised the doubt, shared by many physiologists and high-altitude physicians, that excessive polycythemia may be maladaptive and that an optimal hematocrit exists for a given altitude. Moreover recent sea-level studies demonstrate reduced cerebral blood flow when the hematocrit exceeds 45%. Such findings are the basis for the popular concept that excessive polycythemia in sojourners also may be maladaptive (26).

In this chapter the mechanisms of erythropoiesis are briefly reviewed. Then the practice of bloodletting, or phlebotomy, in high-altitude natives and sea-level sojourners is discussed.

MECHANISM OF ERYTHROPOIESIS

Humoral regulation of the red cell mass, although suspected for many years, was not proven until 1948, when Erslev (10) found clear evidence in cross-circulation experiments. The agent, a glycoprotein, was named erythropoietin (EPO). Jacobson and co-workers (21) in 1957 showed that the kidney is responsible for the bulk of its production; EPO has now been extensively purified. Whether or not renal EPO is the sole regulator of red cell production has been a controversial issue. In adult rats at least 10% of the total EPO may be produced in the liver (12); EPO is produced also in the carotid bodies of cats (43). Recent studies show that EPO can also be released from alveolar and splenic macrophages that have been exposed to silica (37). However, it is not known whether these cells actually produce the hormone or merely serve as storage sites.

Human studies of extrarenal EPO production are of course more difficult and rely mainly on clinical observation. Anephric subjects still maintain a level of erythropoiesis, albeit lower than normal humans (13). However, such subjects can increase EPO production in response to hypoxia (33) or hemorrhage (32). The site of this EPO production is not established, but human liver cells can produce EPO, as shown in patients with hepatocellular carcinoma (27). Indeed many cell types may have the potential to produce this hormone.

Although EPO probably is the principal controller of the rate of erythropoiesis, other factors can modulate its effects. Halvorsen (19) argued that the central nervous system plays a role in erythropoiesis because direct stimulation of the hypothalamus seems to increase the number of reticulocytes in peripheral blood. These observations, however, may be difficult to interpret, because the hypothalamic-pituitary axis can also be involved in plasma volume regulation, which in turn may affect oxygen delivery. Dietary protein intake (3), environmental contamination (23), and partial pressure of carbon dioxide (46) can also modulate EPO production; the effect of thyroxine on protein synthesis may also play a role (12). In view of the complex interaction of variables, it is not surprising that red cell production varies so much among individuals.

The nature of the renal oxygen sensor has not yet been clearly defined. Subjects who have mutant hemoglobins with a high affinity for oxygen also develop polycythemia; their EPO mechanism is intact, as demonstrated by EPO secretion in response to phlebotomy (2). However, although the plasma concentration of oxygen (partial pressure of oxygen) and the oxygen content of the blood are high, the body senses a lack of oxygen delivery. Presumably this is because the transfer of oxygen from hemoglobin to sensing sites is diminished. The sensing mechanism in the kidney (or other sites) is probably a composite function of available oxygen that includes blood flow, hemoglobin concentration, and oxygen affinity.

One major practical problem in EPO studies has been the lack of a widely available satisfactory assay. For many years the standard method has been a mouse bioassay (34) involving exhypoxic or hypertransfused animals. Test materials (concentrated urine or plasma) are injected into these animals, and then the uptake of radioactive iron is measured. Many animals are needed for each point because their responses vary considerably and because standard curves must be prepared for each assay, requiring hundreds of mice for one set of measurements. Recently radioimmunoassays have been developed (40), but so far no satisfactory commercial assay has been available. This situation should be rectified as monoclonal antibody technology is developed.

ERYTHROPOIESIS AT HIGH ALTITUDE

The mechanism of increased erythropoiesis noted by Viault appears to be the rapid rise in serum EPO after exposure to high altitude. Faura and co-workers (14), using the mouse bioassay, measured urinary EPO in seven lowlanders taken quickly to Morococha (4,520 m). After an initial latent period of 6 h, they noted that EPO rapidly rose to reach a peak in 24 h and then fell to a new plateau (about twice sea-level normal) in 48 h. In a similar study by Abbrecht and Littell (1) of five lowlanders in Colorado, serum EPO reached a peak 48 h after a 2-h ascent to 4,360 m and remained significantly elevated for 9 days. After that, however, the serum EPO was indistinguishable from sea-level normal.

The fall of serum EPO to sea-level control values in lowland sojourners at high altitude is a troublesome observation because it occurs before the hemoglobin concentration changes significantly. Thus feedback interaction between EPO and the red cell mass must not be simple. Possibly erythroid cells are more sensitive to EPO after stimulation. Alternatively the rate of EPO turnover could be increased after stimulation of erythroid cells. Perhaps, however, there are other unknown components of the control system. Finally, the bioassay is not well suited to measurements of near-normal values. Immunological techniques that quantitate antigenic material (rather than biological activity) may provide better data.

Clearly highland natives of Peru also have elevated EPO. Faura and co-workers (14) found elevated urinary EPO at Morococha, and Reynafarje and co-workers (36) found that plasma from such subjects accelerated the uptake of radioactive iron when injected into rats. Merino (25), in a most extraordinary experiment, injected pooled plasma from natives of Morococha into volunteers in Lima and showed a clear reticulocytosis compared to control recipients of normal plasma. This observation shows unequivocally that plasma from natives of high altitude contains an erythropoietic factor. Whether hemoglobin concentration and the concentration of this factor (presumed to be EPO) are related in high-altitude natives is undetermined.

The characterization of the erythropoietic mechanism at high altitude is complicated somewhat by studies with high-altitude natives taken suddenly to sea level. Reynafarje et al. (36) found that plasma from such natives depressed iron utilization when injected into rats. Faura et al. (14) found similar results by using a different EPO assay system. These studies suggest the presence of an inhibitor of erythropoiesis. However, similar effects have been observed in subjects with very low EPO concentrations (4) and are probably due to unspecific properties of serum usually masked by EPO. The presence or absence of an EPO antagonist could be crucial for understanding the etiology of CMS. It remains to be seen whether such an inhibitor does exist or if the results are due to an unspecific serum effect. Perhaps the newer assay techniques for EPO will help clarify this important point.

PHYSIOLOGICAL EFFECTS OF POLYCYTHEMIA

Polycythemia is a two-edged sword: increased hemoglobin concentration is potentially important in augmenting oxygen-carrying capacity, but hematocrit values over 50–55% dramatically increase blood viscosity (Fig. 1). The data in Figure 1 were obtained by in vitro viscometry. They do not necessarily correlate quantitatively with blood flow, because of alterations in the peripheral vascular resistance (17), which can partially offset the effects of increased viscosity, particularly during exercise. Nevertheless polycythemia increases resistance to blood flow in both the pulmonary and the systemic circulations and decreases cardiac output (18). However, increases in red cell mass are normally accompanied by increased total blood volume; this change leads to increased venous return and a compensatory increase in cardiac output (6). Whether or not these opposing mechanisms preserve cardiac output in sea-level residents with severe anemia or polycythemia is not known, but they probably do not. Still less is known about these mechanisms in high-altitude natives.

Two studies directly relate to the in vivo effects of variation in hematocrit on oxygen delivery. Cerretelli (7) found that polycythemic climbers at high altitude had reduced maximal oxygen consumption even when they breathed

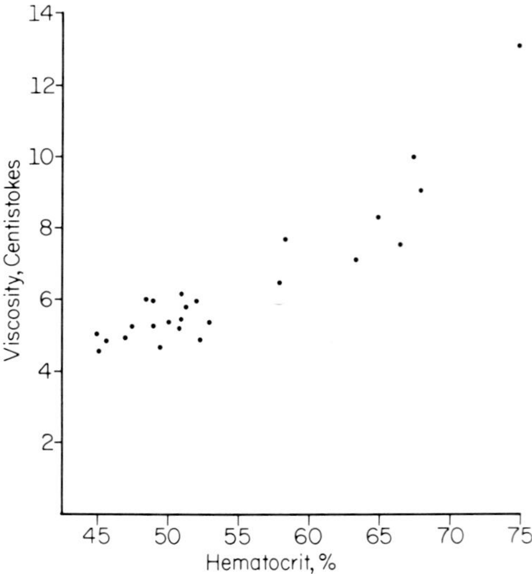

FIG. 1. Relationship between whole blood viscosity and hematocrit. Relationship is approximately exponential, with large increases in viscosity occurring at hematocrits above 50–55%.

100% oxygen. This indicates that peripheral extraction of oxygen from blood was limited by its reduced flow. In contrast Buick et al. (5) showed a slight increase in maximal oxygen consumption in highly trained sea-level athletes when given autologous red cell transfusions. This study indicates that, when hematocrit is low (or normal), peripheral oxygen extraction may be limited by tissue or cardiac factors and not by properties of the blood itself.

CHRONIC MOUNTAIN SICKNESS

The distribution of hematocrit values in the Peruvian Andes is skewed to high values, as typified by natives of Cerro de Pasco (4,250 m) (Fig. 2). Those individuals with excessively high hematocrits suffer with symptoms of plethora. The condition, first described by C. M. Monge (30) in 1928, was originally considered a unique clinical syndrome resulting from loss of acclimatization to altitude. Chronic mountain sickness is still not completely understood, but Whittembury and C. C. Monge (44) suggested that it results from decreased ventilation that normally accompanies aging. They correlated hematocrit with age in a longitudinal study of Andean men over a 15-yr period (Fig. 3). Other workers have not been able to demonstrate correlations between age and hematocrit in high-altitude populations (8), but C. C. Monge and Whittembury (29) stress that this is because individual variation in hematocrit is so great that the changes are obscured in simple mass plots.

The few measurements of red cell mass and plasma volume that have been reported in CMS subjects show only red cell mass to be expanded (20, 38). Penaloza and Sime (35) noted pulmonary hypertension and likened CMS to chronic cor pulmonale or brisket disease in cattle. Ventilation during sleep may also be disordered, as noted by Kryger and co-workers (24), who observed accentuated periods of hypoventilation and arterial desaturation during sleep in CMS subjects in Leadville, Colorado. Whether these conditions are causes or effects of polycythemia needs to be shown clearly.

One remarkable aspect of CMS is its apparent rarity in the Himalayas: no cases are described in the Western literature. There has been much speculation about this, including the hypothesis that South American Indians

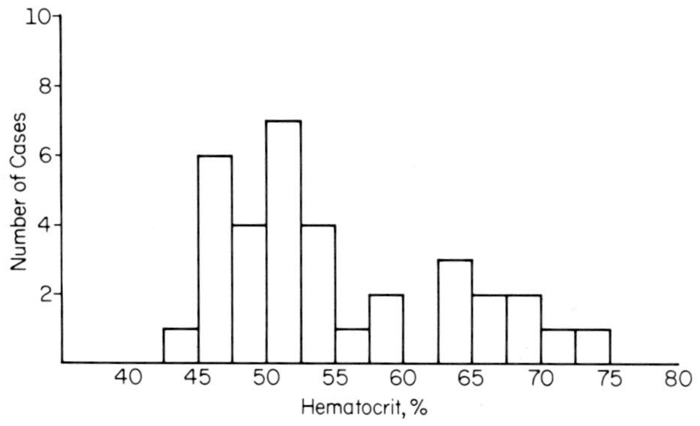

FIG. 2. Distribution of hematocrits in Cerro de Pasco, Peru (4,250 m).

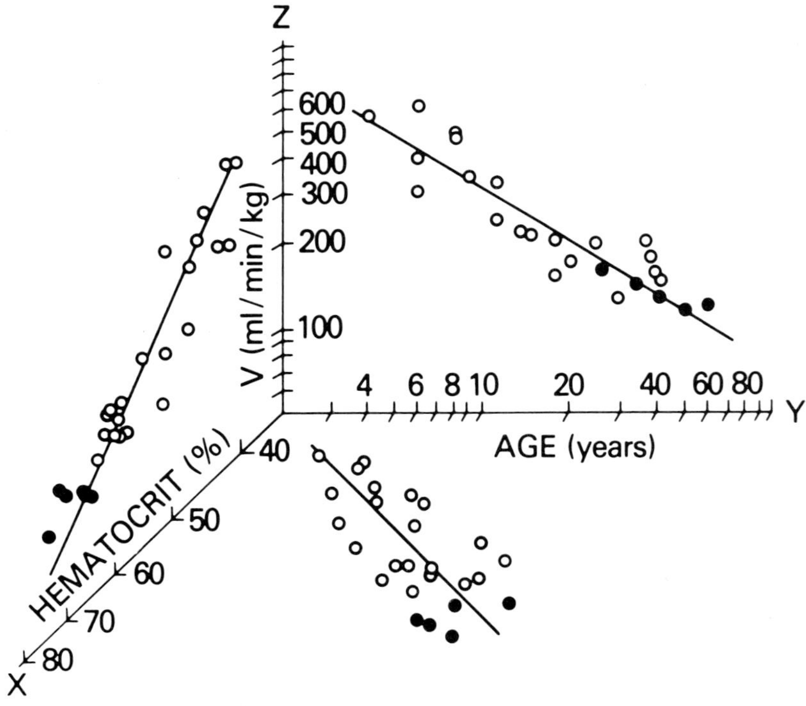

FIG. 3. Relationship between hematocrit, age, and ventilation (V) in high-altitude natives. ●, Chronic mountain sickness values. [From Monge and Whittembury (29).]

have lived at high altitude for only 20,000–30,000 yr, whereas the high Tibetan Plateau has been populated for at least 500,000 yr (29). Thus perhaps insufficient time has been allowed for evolutionary adaptation in the Himalayan natives. However, it is also possible that the two regions differ in other environmental factors or that the populations have different diets or migratory habits. Moreover Himalayan natives may not have been studied in as much detail as their Andean counterparts. I observed one young high-altitude Sherpa on the 1981 American Medical Research Expedition to Everest who had a hematocrit of 72% and typical symptoms of CMS, which he had suffered for several years. However, a diagnosis of CMS requires exclusion of pulmonary disease or other causes of secondary polycythemia, which would require extensive clinical evaluation.

Geographic differences between the two areas could also be important. The Andean range is situated on a continental divide; the Himalayan range is not. Therefore the rivers of the Himalayas have cut deep gorges between the mountains so that most of the passes are at lower altitudes than those of the Andes (15). This means that permanent settlements can be established at lower altitudes in the Himalayas even though the mountains are much higher. Further comparative studies of these two areas could lead to more information of great interest.

PHLEBOTOMY IN HIGH-ALTITUDE NATIVES

It is well known in Andean villages that descent to sea level or phlebotomy relieves the symptoms of CMS. Indeed it is common practice to remove blood from surgical candidates to reduce the risk of excessive bleeding (unpublished observations). However, there are remarkably few data in the scientific literature to document the physiological benefits of bloodletting. In 1965 C. C. Monge et al. (28) reported studies on three subjects in Cerro de Pasco, Peru (4,250 m); small amounts of blood were removed (600 ml, 750 ml, and 1,350 ml). They found no change in symptoms, slight reductions in the partial pressure of carbon dioxide in arterial blood, and increases in plasma pH. Arterial oxygen saturation was measured but showed no regular pattern of change. They concluded that the studies were not representative because clinical improvement was generally well known, and they suggested that larger amounts of blood should have been taken.

Cruz and co-workers (9) reported phlebotomy studies in four subjects with CMS at L'Oroya, Peru (3,700 m). One had atrial fibrillation; another had right bundle branch block and signs of heart failure. The authors demonstrated a small but significant improvement in static pulmonary function after phlebotomy. Our measurements at Cerro de Pasco (unpublished) failed to confirm these results, but some improvement is expected because reduction of blood volume should increase lung capacity (16).

PHLEBOTOMY IN SOJOURNERS TO HIGH ALTITUDE

Healthy lowlanders at high altitudes are susceptible to several characteristic problems (pulmonary infarction, retinal hemorrhage, and frostbite) that could be aggravated by increased blood viscosity. Furthermore Cerretelli (7) has suggested that one reason for decreased performance after long periods (weeks to months) at extreme altitude is decreased tissue extraction of oxygen due to reduced blood flow that accompanies polycythemia.

Many of these problems faced by sojourners are similar to sea-level problems of peripheral vascular disease. This probably explains why hemodilution in mountaineering expeditions has recently become popular, particularly among the Germans, the Swiss, and the Italians. Reports in the surgical literature more and more often advocate the use of hemodilution in patients with peripheral vascular disease (26). These are supported by physiological evidence that cerebral blood flow is exquisitely sensitive to hematocrit variations and that the cardiac oxygen extraction can be maintained over a broad range of arterial oxygen content (22).

The benefits and dangers of phlebotomy or hemodilution in climbers, however, are difficult to establish by experimentation because of the logistics required for the studies. The procedures carry risks, and the idea of suddenly reducing the blood volume by phlebotomy is not appealing to highly motivated healthy climbers. Hemodilution may be safer but requires intravenous infusion of sterile isotonic solutions that must be transported to remote areas. Few expeditions are prepared to cope with hypersensitivity reactions to intravenous solutions or with septicemia.

Nevertheless Zink et al. (45) presented results of uncontrolled hemodilution experiments in climbers on a major Himalayan expedition. He reported that the climbers with diluted blood were less symptomatic and that one of them climbed to 8,500 m without supplemental oxygen. On the 1981 American Medical Research Expedition to Everest, four climbers underwent hemodilution at the end of the expedition (39). However, the hematocrit changes were small; one subject suffered a hypersensitivity reaction after the procedure. On the 1981 Italian-Swiss expedition to Lhotse, an attempt was made to reduce hematocrits by oral hydration; subjects were evaluated after the expedition (P. Cerretelli, unpublished observation).

These examples demonstrate the difficulties in obtaining useful information about hematocrit reduction in sojourners exposed acutely to high altitude. This problem cannot be solved until careful studies in low-pressure chambers are done or until ingenious methods to evaluate the effects of hemodilution or phlebotomy in field studies are developed.

OPTIMAL HEMATOCRIT

Physiological and excessive polycythemia must be distinguished. Figure 2 shows that the distribution of hematocrits in Cerro de Pasco is clearly bimodal; we speculate that the main peak (50–55%) represents the physiological hematocrit for that altitude and that the secondary peak (~65%) is excessive because persons with hematocrits of 65% are usually symptomatic. The optimal hematocrit for a person at a given altitude cannot yet be predicted. A theoretical analysis of the problem by C. C. Monge and Whittembury (29) suggested that the optimum for natives of 4,250 m may be as low as 34%. However, it must be cautioned that physical training, age, prior exposure to high altitude, and period of residence (acclimatization) could all influence the value for a person (31, 41).

An additional important distinction must be made between sojourners and high-altitude natives. C. M. Monge (31) believed that true acclimatization requires years of residence at high altitude; generalizations from observations of natives may not apply to climbers and trekkers. For example, lowlanders of intermediate levels of physical training are found in ever-increasing numbers among trekkers to the Himalayas and may represent a quite distinct group. Furthermore lowlanders briefly exposed to very high altitudes who become the subjects of physiological studies are often fit athletes, in contrast to the wide range of physical conditioning found in permanent high-altitude residents.

High-altitude exposure may be similar to physical training. Marathon runners sometimes prefer to train at moderate altitude because of subjective evidence that it is more effective than training at sea level. Although there are no firm data to support this preference, exposure to altitude may stimulate tissue adaptation: capillary proliferation and increased efficiency of oxygen utilization by muscle may occur in high-altitude natives (42). If so then a hematocrit higher than that found at sea level may be beneficial, just as "blood doping" may be of some benefit to highly trained sea-level athletes (5).

Many questions about erythropoiesis at high altitude remain unanswered. For example, what is the etiology of CMS? Is this interesting disease an example of a normal control mechanism gone awry? What is the optimal hematocrit for an individual under a given set of conditions? How do the various physiological variables involved in oxygen transport interact in humans? How may they be modified by age, disease, or acclimatization? Are there therapeutic measures that can modify oxygen transport to benefit the individual? These and many other questions will certainly be answered now that high-altitude areas are more accessible and performance of sophisticated measurements is possible, particularly with newly emerging electronic technology.

This work was supported in part by National Science Foundation Interagency Agreements INT-77-21795 and INT-80-07728.

The use of tradenames is for identification only and does not constitute endorsement by the Public Health Service or by the United States Department of Health and Human Services.

REFERENCES

1. ABBRECHT, P. H., AND J. K. LITTELL. Plasma erythropoietin in men and mice during acclimatization to different altitudes. *J. Appl. Physiol.* 32: 54–58, 1972.
2. ADAMSON, J. W., A. HAYASHI, G. STAMATOYANNOPOULOS, AND W. F. BURGER. Erythrocyte function and marrow regulation in hemoglobin Bethesda (B145 Histidine). *J. Clin. Invest.* 51: 2883–2888, 1972.
3. ALIPPI, R. M., J. M. GIGLIO, A. C. BARCELO, C. E. BOZZINI, R. FARINA, AND M. E. RIO. Influence of dietary protein concentration and quality on response to erythropoietin in the polycythaemic rat. *Br. J. Haematol.* 43: 451–456, 1979.
4. BIRGEGARD, G., O. MILLER, J. CARO, AND A. ERSLEV. Serum erythropoietin levels by radioimmunoassay in polycythemia. *Scand. J. Haematol.* 29: 161–167, 1982.
5. BUICK, F. J., N. GLEDHILL, A. B. FROESE, L. SPRIET, AND E. C. MEYERS. Effect of induced erythrocythemia on aerobic work capacity. *J. Appl. Physiol.: Respirat. Environ. Exercise Physiol.* 48: 636–642, 1980.
6. CASTLE, W. B., AND J. H. JANDL. Blood viscosity and blood volume: opposing influences upon oxygen transport in polycythemia. *Semin. Hematol.* 3: 193–198, 1966.
7. CERRETELLI, P. Oxygen transport on Mount Everest: the effects of increased hematocrit on maximal O_2 transport. *Adv. Exp. Med. Biol.* 75: 113–119, 1976.
8. CHIODI, H. Aging and high-altitude polycythemia. *J. Appl. Physiol.: Respirat. Environ. Exercise Physiol.* 45: 1019–1020, 1978.
9. CRUZ, J. C., C. DIAZ, E. MARTICORENA, AND V. HILARIO. Phlebotomy improves pulmonary gas exchange in chronic mountain sickness. *Respiration* 38: 305–313, 1979.
10. ERSLEV, A. J. Humoral regulation of red cell production. *Blood* 8: 349, 1953.
11. ERSLEV, A. J. Erythroid adaptation to altitude. *Blood Cells* 7: 495–508, 1981.
12. ERSLEV, A. J., J. CARO, K. SILVER, AND O. MILLER. The biogenesis of erythropoietin. *Exp. Hematol.* 8, Suppl.: 1–13, 1981.
13. ERSLEV, A. J., P. J. McKENNA, J. P. COPELLI, R. J. HAMBURGER, H. E. COHN, AND J. E. CLARK. The rate of red cell production in two nephrectomized patients. *Arch. Intern. Med.* 122: 230–235, 1968.
14. FAURA, J., J. RAMOS, C. REYNAFARJE, E. ENGLISH, P. FINNE, AND C. A. FINCH. Effect of altitude on erythropoiesis. *Blood* 33: 668–676, 1969.
15. GILBERT, D. L. The first documented report of mountain sickness: the China or headache mountain story. *Respir. Physiol.* 52: 315–326, 1983.
16. GLASER, E. M., AND J. McMICHAEL. Effect of venesection on the capacity of the lungs. *Lancet* 2: 230–231, 1940.
17. GUSTAFSSON, L., L. APPELGREN, AND H. E. MYRVOLD. Polycythemia, viscosity and blood flow in working and non-working skeletal muscle in the dog. *Bibl. Anat.* 18: 56–59, 1979.
18. GUYTON, A. C., C. E. JONES, AND T. G. COLEMAN. *Cardiac Output and Its Regulation* (2nd ed). Philadelphia, PA: Saunders, 1973.
19. HALVORSEN, S. The central nervous system in regulation of erythropoiesis. *Acta Haematol.* 35: 65–79, 1966.
20. HURTADO, A., C. F. MERINO, AND D. DELGADO. Influence of anoxemia on erythropoietic activity. *Arch. Intern. Med.* 75: 284–323, 1945.
21. JACOBSON, L. O., E. GOLDWASSER, W. FREED, AND L. PLZAK. Role of the kidney in erythropoiesis. *Nature London* 179: 633, 1957.
22. JAN, K.-M., AND S. CHIEN. Effect of hematocrit variations on coronary hemodynamics and oxygen utilization. *Am. J. Physiol.* 233 (*Heart Circ. Physiol.* 2): H106–H113, 1977.
23. KOLLER, L. D., J. H. EXON, AND J. E. NIXON. Poly-

cythemia produced in rats by environmental contaminants. *Arch. Environ. Health* 34: 252–255, 1979.
24. KRYGER, M., R. McCULLOUGH, R. DOEKEL, D. COLLINS, J. V. WEIL, AND R. F. GROVER. Excessive polycythemia of high altitude: role of ventilatory drive and lung disease. *Am. Rev. Respir. Dis.* 118: 659–665, 1978.
25. MERINO, C. F. *The Plasma Erythropoietic Factor in the Polycythemia of High Altitudes.* School of Aviation Medicine, United States Air Force, Randolph Base, Texas, Rep. 56, November 1956.
26. MESSMER, K. Hemodilution. *Surg. Clin. North Am.* 55: 659–678, 1975.
27. MIRAND, E. A., AND G. D. MURPHY. Erythropoietin alterations in human liver disease. *NY State J. Med.* 71: 860–864, 1971.
28. MONGE, C. C., R. LOZANO, AND J. WHITTEMBURY. Effect of blood-letting on chronic mountain sickness. *Nature London* 107: 770, 1965.
29. MONGE, C. C., AND J. WHITTEMBURY. Chronic mountain sickness and the physiopathology of hypoxemic polycythemia. In: *Hypoxia: Man at Altitude*, edited by J. R. Sutton, N. L. Jones, and C. S. Houston. New York: Thieme-Stratton, 1982, p. 51–56.
30. MONGE, C. M. La enfermedad de los Andes. Sindromes eritremicos. *An. Fac. Med. Lima* 11: 1, 1928.
31. MONGE, C. M. Life in the Andes and chronic mountain sickness. *Science* 95: 79–84, 1942.
32. NATES, J. P., AND M. WITTER. Presence of erythropoietin in the plasma of one anephric patient. *Blood* 31: 249–251, 1968.
33. NATHAN, D. G., E. SCHUPAK, F. STAHLMAN, AND J. P. MERRILL. Erythropoiesis in anephric man. *J. Clin. Invest.* 43: 2158–2165, 1964.
34. NCCLS Proposed standard: PSH-6 Standard Assay for the Determination of Erythropoietin Activity in Body Fluids. 1979.
35. PENALOZA, D., AND F. SIME. Chronic cor pulmonale due to loss of altitude acclimatization (chronic mountain sickness). *Am. J. Med.* 50: 728–743, 1971.
36. REYNAFARJE, C., J. RAMOS, J. FAURA, AND D. VILLAVICENCIO. Humoral control of erythropoietic activity in man during and after altitude exposure. *Proc. Soc. Exp. Biol. Med.* 116: 649–650, 1964.
37. RICH, I. N., AND B. KUBANEK. Release of erythropoietin from macrophages mediated by phagocytosis of crystalline silica. *J. Reticuloendotheliol. Soc.* 31: 17–30, 1980.
38. SÁNCHEZ, C., C. MERINO, AND M. FIGALLO. Simultaneous measurement of plasma volume and cell mass in polycythemia of high altitude. *J. Appl. Physiol.* 28: 775–778, 1970.
39. SARNQUIST, F. H., R. B. SCHOENE, AND P. H. HACKETT. Exercise tolerance and cerebral function after acute hemodilution of polycythemic mountain climbers. *Physiologist* 25: 327, 1982.
40. SHERWOOD, J. B., AND E. GOLDWASSER. A radioimmunoassay for erythropoietin. *Blood* 54: 885–893, 1979.
41. SMITH, E. E., AND J. W. CROWELL. Role of an increased hematocrit in altitude acclimatization. *Aerospace Med.* 22: 39–43, 1967.
42. TENNEY, S. M., AND L. C. OU. Physiological evidence for increased tissue capillarity in rats acclimatized to high altitude. *Respir. Physiol.* 8: 137–150, 1970.
43. TRAMEZZANI, J. H., E. MONITA, AND S. R. CHICCHIO. The carotid body as a neuroendocrine organ involved in control of erythropoiesis. *Proc. Natl. Acad. Sci.* 68: 52–55, 1971.
44. WHITTEMBURY, J., AND C. C. MONGE. High altitude, hematocrit, and age. *Nature London* 238: 278–279, 1972.
45. ZINK, R. A., W. SCHAFFERT, W. BRENDEL, K. MESSMER, E. SCHMIDT, AND P. BRENETT. Hemodilution in high altitude mountain climbing: a method to prevent or treat frostbite, high altitude pulmonary edema, and retinal hemorrhage. *Abstracts of Scientific Papers, Am. Soc. Anesthesiolog. Annu. Meet., Chicago, 1978,* p. 93.
46. ZUCALI, J. R., LEE, M., AND MIRAND, E. A. Carbon dioxide effects on erythropoietin and erythropoiesis. *J. Lab. Clin. Med.* 92: 648–655, 1978.

15

Ventilatory Function in Adaptation to High Altitude: Studies in Tibet

S. Y. HUANG, X. H. NING, Z. N. ZHOU,
Z. Z. GU, AND S. T. HU

*Shanghai Institute of Physiology, Academia Sinica,
Shanghai, China*

HUMAN BEINGS HAVE LIVED at high altitudes on the Tibetan Plateau for thousands of years. The Tibetans, having perhaps lived at high altitude longer than any other population on earth, have had ample opportunity to adapt. One important adaptation is ventilation. Pulmonary ventilation is the first step in the transport chain that moves oxygen from the ambient air to the tissue cell. Failure of pulmonary ventilation in individuals who go to high altitude is considered central in acute and chronic altitude sickness. We hypothesized that the healthy Tibetan at high altitude represents the "gold standard" of ventilatory acclimatization. Newcomers who successfully adapt to life on the Tibetan Plateau might show ventilatory function as good as that of the native Tibetan, and those who show maladaptation show poor ventilatory function.

To test this hypothesis we compared ventilatory function in acclimatized sojourners and Tibetan highlanders. The sojourners were from lowland China and had resided at 5,000 m for 1 mo prior to the study. They were the climbers and members of the scientific team of the 1966 Mount Qomolungma (Mount Everest) Scientific and Mountaineering Expedition. All studies were conducted at Base Camp (5,000 m). All subjects were fit and felt well during the 1-mo acclimatization period. The Tibetans were also climbing and support members of the expedition. They had lived on the Tibetan Plateau (altitudes from 3,000 to 4,500 m) from birth and had also resided at 5,000 m for 1 mo prior to study. The groups were of comparable sex (all male), age, and body size (Table 1).

Measurements of minute ventilation ($\dot{V}_E$) were obtained with a standard mouthpiece and noseclip and a dry-gas meter. The collection period was at least 5 min. Respiratory frequency (f) was counted with a stopwatch; tidal volume (V_T) was also calculated. Alveolar gases were collected in a Haldane sampler and were analyzed for oxygen and carbon dioxide with a Haldane gas analyzer.

The results indicated similar $\dot{V}_E$ (Fig. 1), f, and V_T (Fig. 2) values among the acclimatized sojourners and the Tibetan highlanders. The alveolar carbon dioxide ($P_{A_{CO_2}}$) and oxygen ($P_{A_{O_2}}$) tensions were also similar (Fig. 3). These results confirmed the working hypothesis that ventilation in well-acclimatized lowlanders was similar to that of the native Tibetans. However, the lowlander

TABLE 1. *Characteristics of male subjects studied after 1 mo at 5,000 m in Tibet*

	n	Age, yr	Height, cm	Weight, kg
Acclimatized Sojourners	48	29	169	60
Tibetan Highlanders	91	24	167	62

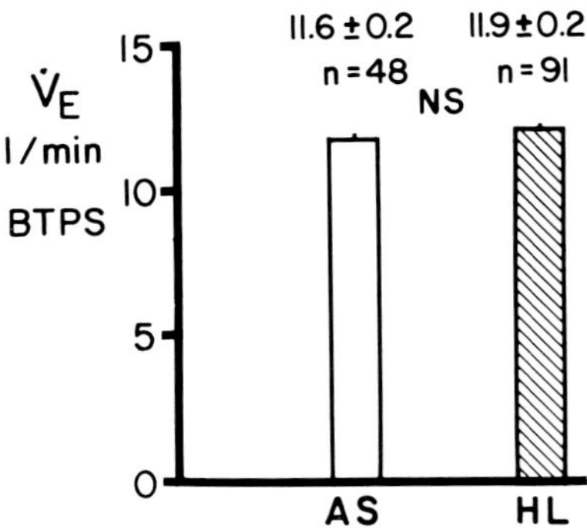

FIG. 1. Similar minute ventilation ($\dot{V}_E$) in acclimatized sojourners (AS) and in Tibetan highlanders (HL) at 5,000 m (mean ± SE). *n*, No. of subjects; NS, not significant ($P > 0.05$).

cohort contained fit athletes, and the duration of high-altitude residence (1 mo) was short. Thus in 1972 we made similar measurements in 20 healthy male government workers from lowland China who had resided for at least 1 yr in Xigaze at 3,890 m. Their ages ranged from 20 to 24 yr. Twelve male Tibetans were studied for comparison. They were also government workers residing at altitudes above 3,500 m from birth, and they also ranged in age from 20 to 24 yr. In addition to ventilation, measurements of arterial P_{CO_2} (Pa_{CO_2}) and arterial pH were obtained in arterialized venous blood from the heated hand (45°C) analyzed by a Radiometer gas analyzer. Arterial oxygen saturation (Sa_{O_2}) in the arterialized blood was calculated by dividing oxygen content by oxygen capacity, as measured by the Van Slyke method. Carbon monoxide diffusing capacity of the lung (DL_{CO}) was measured with a steady-state method (Godart diffusion test). The results indicate no significant differences between the two groups in any of the measurements (Table 2). Thus healthy nonathletes who had migrated from low altitude and had resided for at least 1 yr on the Tibetan Plateau had ventilatory measurements similar to native Tibetans.

If ventilation is important in the acclimatization process, then those persons who fail to adapt well to life on the Tibetan Plateau might show impairment in their ventilatory measurements. To approach this aspect of the hypothesis we made ventilatory measurements in persons who had adapted poorly to high altitude. Their measurements were compared with those from low-altitude persons who had adapted well to high altitude. This comparison was a particularly convenient experimental design, and, as we have shown, well-acclimatized sojourners have ventilatory measurements similar to those of the Tibetans. This approach necessitated a definition of well-adapted and poorly adapted sojourners from low altitude in Tibet. The blood erythrocyte count and hemoglobin content were possible indicators. For example, person-

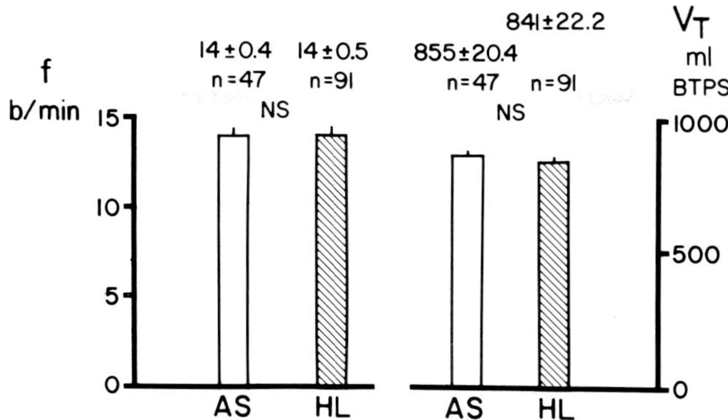

FIG. 2. Similar respiratory frequency (f) and tidal volume (VT) in acclimatized sojourners (AS) and in Tibetan highlanders (HL) at 5,000 m (mean ± SE). *n*, No. of subjects; NS, not significant ($P > 0.05$).

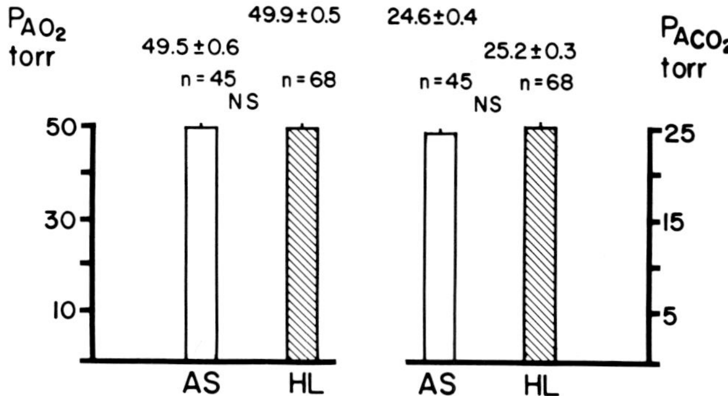

FIG. 3. Similar alveolar O_2 ($P_{A_{O_2}}$) and CO_2 ($P_{A_{CO_2}}$) tensions in acclimatized sojourners (AS) and in Tibetan highlanders (HL) at 5,000 m (mean ± SE). *n*, No. of subjects; NS, not significant ($P > 0.05$).

nel at a local hospital in Tibet surveyed persons living at altitudes of 4,000–4,500 m. They found that 67 of 166 male sojourners from low altitude having routine blood counts had red cell counts over $7 \times 10^6/\text{mm}^3$ and hemoglobin concentrations over 20 g/100 ml. By comparison, only 1 of 37 female sojourners had such abnormal blood values. Among the native Tibetan population the abnormal values were found in 26 of 159 males and in none of 74 females. From these considerations it appeared that excessive polycythemia was more common among male sojourners than among male Tibetans. Excessive polycythemia has frequently been used as a marker of chronic mountain sickness. We thus compared ventilatory measurements in male sojourners who had normal blood values with ventilatory measurements in those with excessive polycythemia.

The study was conducted in 1972 in Xigaze at 3,890 m. Studied were 56 healthy government workers, age 20–24 yr, and 52 government workers with excessive polycythemia. All 108 subjects had moved from lowland China to

TABLE 2. *Ventilatory measurements at 3,890 m*

$\dot{V}_E$, liters/min BTPS	V_T, ml BTPS	f, beats/min	$\dot{V}_A$, liters/min BTPS	Pa_{CO_2}, mmHg	pHa, units	Sa_{O_2}, %	DL_{CO},* ml·min^{-1}·mmHg^{-1}
			Acclimatized sojourners (n = 20)				
15.9 ±0.9	914 ±60	17.5 ±0.9	12.9 ±0.7	28.2 ±0.4	7.419 ±0.004	84.8 ±0.8	25.3 ±1.6
			Tibetan highlanders (n = 12)†				
16.9 ±0.9	1041 ±100	16.7 ±1.3		30.7 ±1.9	7.412 ±0.007	83.1 ±1.6	27.6 ±2.0
			Excessively polycythemic patients (n = 14)‡				
12.5 ±0.8	565 ±13	22.3 ±1.3	9.1 ±0.7	35.5 ±0.7	7.382 ±0.005	77.6 ±1.1	14.4 ±1.1

Values are means ± 1 SE. $\dot{V}_E$, minute ventilation; V_T, tidal volume; f, respiratory frequency; $\dot{V}_A$, alveolar ventilation; Pa_{CO_2}, partial pressure of CO_2 in arterial blood; pHa, arterial pH; Sa_{O_2}, arterial O_2 saturation; DL_{CO}, pulmonary diffusion capacity for CO. *Not corrected for hemoglobin concentration. †Values do not differ significantly ($P > 0.05$) from those in acclimatized sojourners. ‡Values differ significantly from those in healthy subjects ($P < 0.05$).

the Tibetan Plateau more than 1 yr prior to the study. The average blood value among the healthy subjects was 6 × 10^6/mm^3, 17.9 g/100 ml, and 56% for red cell count, hemoglobin concentration, and hematocrit, respectively. The same measurements in the polycythemic subjects were 7.6 × 10^6/mm^3, 22.5 g/100 ml, and 78%. From among these subjects we obtained more complete studies in 20 healthy and 14 polycythemic subjects. The 14 polycythemic subjects had been hospitalized for signs and symptoms of chronic mountain sickness and were about to begin treatment. Chronic mountain sickness was the only disease identified in these 14.

The results indicated that the patients with chronic mountain sickness had lower $\dot{V}_E$, V_T, alveolar ventilation, arterial pH, Sa_{O_2}, and DL_{CO} values (Table 2). The patients also had higher f and Pa_{CO_2} values. These results indicated that the poorly acclimatized subjects who had polycythemia and required hospitalization for chronic mountain disease had impaired ventilatory and gas-exchange functions.

From these measurements we conclude that *1)* polycythemia in Tibetans is less excessive than in sojourners from low altitudes. The Tibetan men are not completely immune to excessive polycythemia, but their relative protection may stem from adaptation over many generations at high altitude. Women on the Tibetan Plateau are less susceptible to polycythemia than men. Polycythemia probably indicates the magnitude of the chronic hypoxic stress suffered by the individual at a particular altitude. It represents some integration of the degrees of hypoxia experienced both at night and during the day and often indicates hypoventilation. *2)* Acclimatized new arrivals to Tibet from low altitudes and those who have lived there for many months and have remained well have ventilatory measurements not different from the healthy Tibetan. *3)* Persons who have lived in Tibet for more than 1 yr and who have chronic mountain sickness have impaired ventilation and gas exchange compared to healthy sojourners and, by implication, to healthy Tibetans. The ventilatory insufficiency in sojourners with chronic mountain sickness might reflect a diminished respiratory sensitivity. Their reduced pulmonary diffusion capacity is probably caused by hypoventilation or a thickened alveolar capillary membrane.

The evidence supports the concept that Tibetans are adapted to high altitude better than persons who come from a low-altitude population. An important part of the adaptation appears to be ventilatory acclimatization.

The authors are very grateful to Dr. J. T. Reeves for his help in the preparation of the manuscript.

REFERENCES

1. HU, S.-T., Z.-Z. GU, X.-H. NING, Z.-N. ZHOU, H.-Y. LIN, Z.-M. ZENG, Z.-Z. CHEN, AND T.-C. PAN. The role of respiratory function in the pathogenesis of severe hypoxemia in chronic mountain sickness. In: *Geological and Ecological Studies of Qinghai-Xizang Plateau.* New York: Gordon & Breach, p. 1427–1433. [Proc. Symp. Qinghai-Xizang (Tibet) Plateau, Beijing, China, vol. 2.]
2. HUANG, S.-Y., Z.-Z. GU, C.-F. PA, AND S.-T. HU. Ventilatory control in Tibetan highlanders. In: *Geological and Ecological Studies of Qinghai-Xizang Plateau.* New York: Gordon & Breach, p. 1363–1369. [Proc. Symp. Qinghai-Xizang (Tibet) Plateau, Beijing, China, vol. 2.]

16

Ventilation in Human Populations Native to High Altitude

PETER H. HACKETT, JOHN T. REEVES, ROBERT F. GROVER, AND JOHN V. WEIL

Department of High Latitude Studies, University of Alaska, Anchorage, Alaska; and Cardiovascular Pulmonary Research Laboratory, University of Colorado Health Sciences Center, Denver, Colorado

PULMONARY VENTILATION MOVES AIR from the atmosphere into the alveoli, where the oxygen in the air enriches the blood. This ventilation is the first link in the chain of oxygen transport from atmosphere to the tissues and cells of the body. At high altitude, where the pressure of oxygen in the atmosphere is less than at sea level, the body compensates by increasing ventilation. The compensation is greater in some persons than in others. Those who breathe more have more oxygen in the arterial blood and less carbon dioxide and might be expected to fare better at high altitude than those who breathe less. The adaptation (increase) in pulmonary ventilation at high altitude may not be complete in days or even weeks. In fact ventilatory adaptation may continue over many generations.

Four high-altitude populations in which ventilation has been examined are the South American Indians, the Sherpas of Nepal, inhabitants of the Tibetan Plateau, and persons living in Leadville, Colorado, North America's highest community (3,100 m). Ventilation in all groups is higher than in the sea-level human, but the degree and duration of hyperventilation may vary among them. When compared to lowlanders recently acclimatized to high altitudes, some high-altitude natives, notably South American, have a lower pulmonary ventilation (5, 21, 27). This has been interpreted as an energy-efficient adaptation (20, 32) because the work of breathing is reduced. Natives of Tibet, however, a population that has probably been at altitude longer than the other populations (and thus may be better adapted), have ventilation similar to that of the well-acclimatized Chinese who had come to live in Tibet (see chapt. 15). The Sherpas, a subpopulation of Tibetans, are renowned for their capacity for physical work at extreme altitudes. However, measurements relating to resting ventilation in Sherpas are limited. A previous study (32) at 4,880 m altitude involved simultaneous sampling of cerebrospinal fluid and end-tidal gas in three Sherpas and five acclimatized westerners. The end-tidal partial pressures of carbon dioxide (P_{CO_2}) (Sherpa: 26.8 mmHg; westerner: 26.2 mmHg) and oxygen (P_{O_2}) (Sherpa: 52.5 mmHg; westerner: 50.5 mmHg) were not different between the groups, but the Sherpas had lower values of pH in their cerebrospinal fluid (7.328 vs. 7.374). The arterial blood of the five westerners and the three Sherpas plus one additional Sherpa was also sampled. On this occasion the end-tidal P_{CO_2} values were higher in the Sherpas than in the westerners (28.6 mmHg vs. 25.9 mmHg). Considering how few Sherpas had been examined, we wished to obtain additional measurements related to ventilation, to examine other components of ventilatory adaptation (e.g., vital capacity, respiratory pattern, and control of ventilation), and to compare these to both western lowlanders and other high-altitude populations.

The first considerations were the height and weight of the Sherpa male under investigation. We wondered if the height-weight relationship for Sherpas was different from that for western males, because differences in body configuration might indicate that ventilatory standards developed for westerners might not apply to Sherpas. In 25 Sherpas and 25 western trekkers measured in Katmandu, the Sherpas were smaller than westerners but the height-weight relationships were similar (Fig. 1; 14). Therefore ventilatory measurements were corrected for the differences in body size so that a comparison could be made.

In our study of the respiratory system, we first considered the chest size and vital capacity (Table 1). The size of the Sherpa chest was reported in one study to be similar to that of the Andean chest (51), which is larger than that of the low-altitude resident (12, 41, 42). However, direct comparison of the Sherpa population to the low-altitude Tibetan and Nepali populations when corrected for differences in body sizes showed no increase in chest size of Sherpas (22, 44), and Pawson (44) concluded that chest size in Sherpas was not increased like in South Americans. Beall (1) compared the chest morphology of a Quechua population to that of a Tibetan population in Nepal at a similar altitude and concluded that chest depth was similar, whereas chest width of Tibetans was smaller (Table 1).

Therefore, whereas American Indians do have large chests for their body size, the issue is not clear in Sherpas. We found no previous reliable measure-

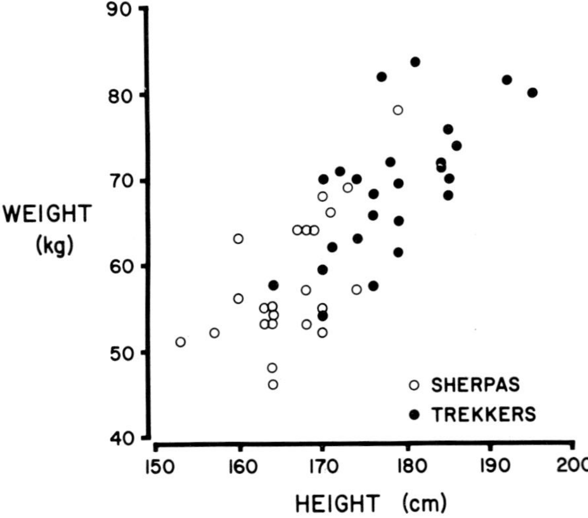

FIG. 1. Weight-height relationship in 25 western trekkers and 25 Sherpas. As shown by linear regression, the groups have similar relationships. [From Hackett et al. (14).]

TABLE 1. *Anthropometric data of high-altitude natives*

	Altitude, m	Age, yr	Height, cm	Weight, kg	Chest measurements, cm*			Ref.
					Circumference	Width	Depth	
Sherpa								
$n = 61$	3,500		162.2	54.6	84.6 (0.52)			42
$n = 25$	2,880	25.8	164.7	51.1	88.5 (0.537)			23
$n = 109$	2,600	18–85	163.1	56.3		28.8 (0.177)	19.7 (0.120)	51
Tibetan								
$n = 28$	3,800	25–35	160.1			26.6 (0.166)	20.4 (0.127)	1
Andean								
$n = 40$	4,000	30	158.9	55.2	89.7 (0.565)	27.6 (0.174)	20.6 (0.129)	12
$n = 52$		25–35	160.0			28.2 (0.176)	20.5 (0.128)	1

Lahiri et al. (29) reported that Andean children 9–13 yr old at 3,850 m had vital capacities of 2.87 ± 0.62 liters, which was 121 ± 14% of predicted. Young adults 18–21 yr old had vital capacities of 5.05 ± 0.67 liters (145% of predicted), whereas the values for sea-level controls were 3.73 ± 31 liters (104% of predicted). *n*, No. of subjects. *Values in parentheses are chest measurements divided by height.

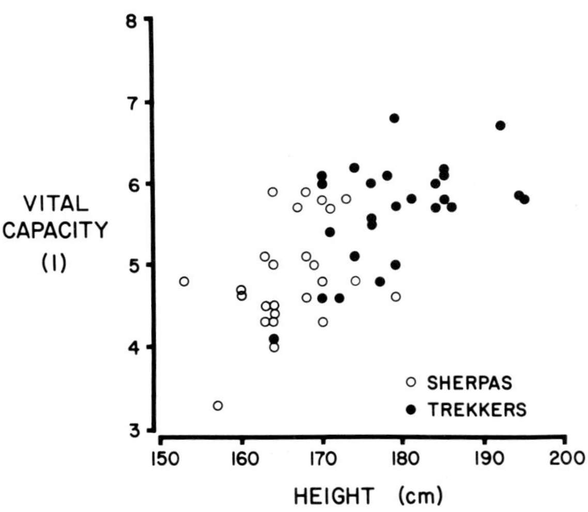

FIG. 2. Relationship of vital capacity to height is similar in western trekkers and Sherpas. [From Hackett et al. (14).]

ments of vital capacity in Sherpas. Our results (14), expressed as mean percent of predicted value (±SE) by using an equation for Caucasians based on height, age, and sex, indicated that forced vital capacities are not larger than those of trekkers (Sherpa: 102.7 ± 2.6%; trekker: 104.5 ± 2.2%). Furthermore the relationship of vital capacity to height was similar in Sherpas and trekkers (Fig. 2). The vital capacities of Sherpas did not appear to be increased for their body size. In contrast, high-altitude Peruvians showed vital capacities 6–45% greater than low-altitude natives (20, 29, 56). If the Sherpa does have a large chest in relation to height but not a large increase in vital capacity then residual volume may be increased, as it is increased (by 50%) in Peruvians (56), and more surface area would be available for gas exchange. Measurements of total lung capacity and diffusing capacity in Sherpas are necessary to confirm or refute this suggestion.

We next consider the actual amount of air moved by the lungs. The resting minute ventilation ($\dot{V}_E$) in our study (14) was greater in Sherpas than in trekkers, both at low and high altitudes [low altitude: 7.37 ± 0.34 vs. 5.94 ± 0.37 liters·min^{-1}·m^{-2} (means ± SE, $P < 0.05$); high altitude: 9.8 ± 1 vs. 7.8 ± 0.47 (means ± SE, $P = 0.05$)]. Our data indicated that the higher $\dot{V}_E$ in Sherpas was due to an increased respiratory frequency (low altitude: 20 ± 1 vs. 14 ± 0.7; high altitude: 17 ± 1 vs. 15 ± 1) more than an increased tidal volume. The relative hyperventilation was supported by an increased alveolar P_{O_2} ($P_{A_{O_2}}$) and decreased alveolar P_{CO_2} ($P_{A_{CO_2}}$) at low altitude and an increased arterial oxygen saturation (Sa_{O_2}) at high altitude (14). The results suggested that the Sherpa high-altitude natives ventilated more at low altitude than lowlanders. High-altitude natives in North and South America brought to low altitude have variously been shown to either hyperventilate or hypoventilate compared to lowland natives (8, 49, 50, 53, 55, 56). At high altitude we found in eight Sherpas at least as much ventilation at rest as in acclimatized lowlanders at high altitude (14). However, more measurements of $\dot{V}_E$ and end-tidal or blood P_{CO_2} are needed to evaluate the ventilation in Sherpas at low and high altitude. In chapter 15, Huang et al. report similar ventilation and arterial blood gases in Tibetan highlanders and acclimatized Chinese newcomers to Tibet. However, relative hypoventilation has been reported in Peruvian natives at high altitude (5, 27, 31, 49, 56). A direct comparison between Himalayan and Andean natives is fraught with difficulty because of the differences in altitudes of residence, age, and methods and because of the small sizes of the groups.

The ventilation data in Sherpas during exercise is also limited. Initially Lahiri (32) reported on four Sherpas, and Pugh (45) reported on one Sherpa. Subsequently Lahiri et al. (34) reported on four Sherpas at 2,900 m and eight at 3,800 m. These studies indicated that Sherpas had relative hypoventilation for a given level of exercise with lower $\dot{V}_E$, lower oxygen ventilatory equivalents ($\dot{V}_E/\dot{V}_{O_2}$), higher $P_{A_{CO_2}}$, and lower arterial pH. However, in some of these studies the Sherpas may have had higher maximum oxygen uptakes than the westerners. Differences in ventilation related to oxygen uptake make it difficult to interpret chemosensitivity when the groups have different aerobic capacities. In a subsequent study 21 Sherpas showed $\dot{V}_E/\dot{V}_{O_2}$ values identical to acclimatized mountaineers until maximum oxygen consumption was approached, when ventilation increased in the lowlanders (4). Interestingly high school athletes born and raised in Leadville, Colorado, hyperventilated to the same degree as newcomers at altitude during exercise and did not display the relative hypoventilation reported in Andean natives (13). Thus how the exercise ventilation of the Sherpa highlander relates to ventilation of other persons is not clear. One problem is that total ventilation is a crude measure. As parameters, alveolar ventilation and arterial P_{CO_2} (Pa_{CO_2}) are more sensitive.

One control of ventilation is the hypoxic ventilatory response (HVR). Our own data (14) indicate that, although there was considerable scatter in Sherpas and western trekkers, the mean value (±SE) for HVR, measured as parameter

A and corrected for body size (to account for the small size of Sherpas), was similar in the Sherpa and western trekker [40 ± 7.3 vs. 54 ± 6 (P = 0.3)]. However, uncorrected *A* values were significantly lower in the Sherpa [65 ± 12 vs. 101 ± 11 ($P < 0.05$)]. Thus the Sherpa group did not have significantly blunted hypoxic ventilatory drives, which is in contrast to the reports in Leadville residents (57) and Peruvians (30) and a previous study in Sherpas (37) but similar to a report in Tibetan highlanders (19).

Comparisons of hypoxic chemosensitivity between populations are difficult because of genetic (7, 48), racial (9), and body size (17) differences, which affect HVR. Furthermore there is marked scatter of measured responses in a given population (17). The altitudes at which the lowlander controls and the high-altitude natives are tested also must be considered. Many reports do not give corrected measurements for body size. The populations may also differ in their mobility between altitudes. For example, some Sherpas may spend the warm months at higher altitude but migrate to lower altitudes for the winter. Finally, it is difficult to compare results of different techniques for measuring hypoxic chemosensitivity. However, one possible cause of the variation has been identified.

Studies from high altitude in Peru (29) and Leadville (2) show that HVR decreases with duration of residence. For example, in Peruvian high elevations the HVR was maximal in children at age 8 and decreased in older persons. This supports the concept of acquired chemoreceptor attenuation. The HVR is unrelated to age at low altitude (17) except over an extreme range (24). Thus age alone does not blunt the hypoxic ventilatory responsiveness in the groups studied. Perhaps the maintained hypoxia of long residence at high altitude is responsible for the blunting.

Besides the duration of high-altitude exposure, the magnitude of the hypoxic stimulus (altitude of residence) also seems to affect the HVR. Indeed in cats blunting of hypoxic chemosensitivity has only been observed when the hypoxia is severe (54). To account for the severity of the hypoxia and the duration of hypoxic exposure, we have used an index of altitude of residence (in kilometers) times duration of residence (in years), which we related to HVR. The result (Fig. 3) shows that the HVR values of the Sherpas and Leadville residents become blunted with increasing altitude-years values (2). Thus part of the difference between the HVR values in our study of Sherpas and the values in South American and North American populations may be due to the young age of our Sherpa population [age = 28 ± 1 yr (mean ± SE)] and the lower altitudes at which the Sherpas, like the Tibetans, generally live compared to the Peruvians.

Because the Himalayas are much higher than the Andes and because the Sherpas accompany expeditions to these lofty peaks, Sherpa climbers often go to altitudes between 5,000 and 8,000 m for short periods of time. Their main population centers, however, are at 2,500–3,800 m; their highest villages, which are only sparsely populated, are at 4,000–4,800 m, the same altitude of populous cities in the Andes. Higher HVR in our Sherpas when compared to other high-altitude natives may be at least partially explained by less hypoxic stress and

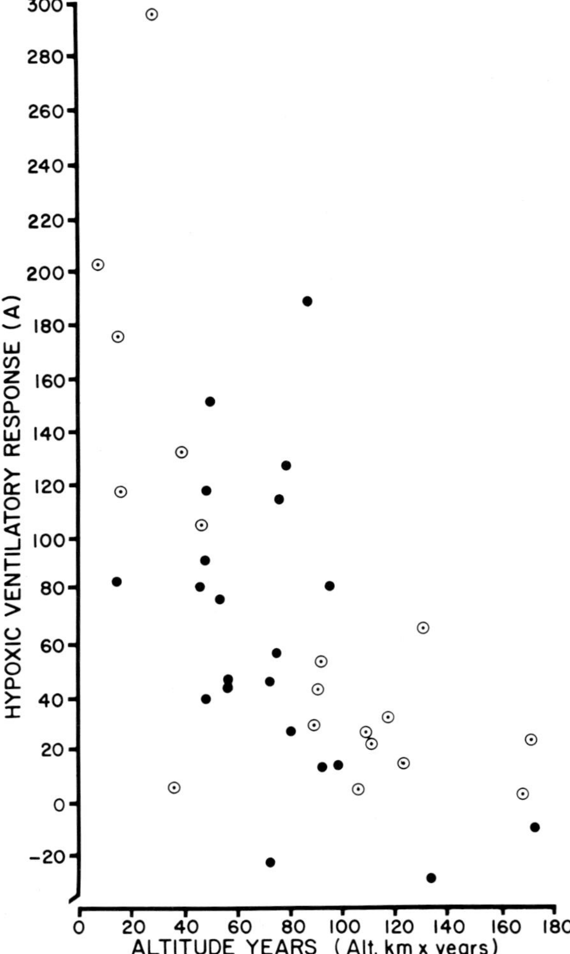

FIG. 3. Effect of altitude years (altitude in kilometers × years at that altitude) on hypoxic ventilatory response as measured by parameter *A* is similar in Sherpas (●) and residents and natives of Leadville, Colorado (⊙). Relationship is significant for all groups. [Adapted from Sørensen and Severinghaus (52).]

fewer years of exposure, resulting in a smaller altitude-years effect. A comparative study of Himalayan and Andean natives with the same techniques and taking into account the altitude-years effect is needed.

What is the significance of a blunted HVR in terms of ventilation? Previous studies have suggested that the change in ventilation of lowlanders when ascending to high altitude is related to the HVR (23, 46). The HVR also appeared to relate to symptoms of acute mountain sickness (Fig. 4; 15). Data from Peru indicate that the relative hypoventilation of those high-altitude natives is associated with a decreased HVR (30, 33, 49, 52). Young Tibetan highlanders (mean age = 24 yr) who did not ventilate less than acclimatized lowlanders also maintained their HVR (19). Studies from Peru have shown that high-altitude natives with blunted drives have relative hypoventilation and increased Pa_{CO_2} at rest for a given altitude, compared to acclimatized lowlanders (5, 27). However, Forster et al. (11) reported that resting ventilation in Leadville natives with blunted drives was slightly lower than acclimatized

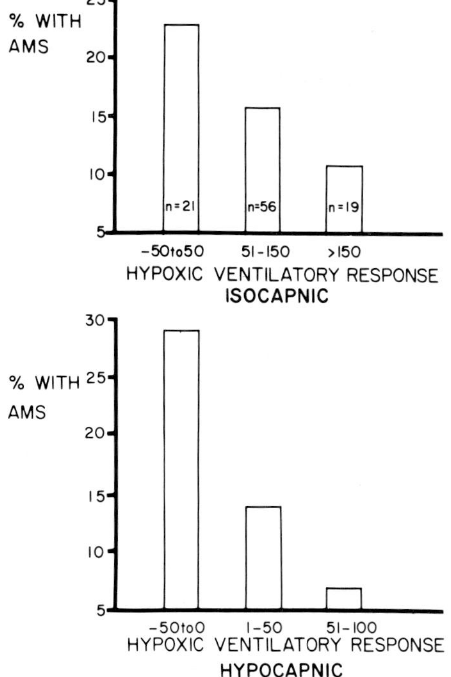

FIG. 4. Percent of subjects with acute mountain sickness (AMS) grouped into low, medium, and high hypoxic ventilatory responses, both isocapnic and hypocapnic. Those with lower hypoxic drives have a higher incidence of AMS. (n = 96, P = 0.1.)

newcomers but that end-tidal carbon dioxide and arterial pH and P_{CO_2} values were not different. Kryger (26) found blunted drives in Leadville natives but hypoventilation only in those with chronic mountain sickness. Thus the relationship of hypoxic drive to ventilation at high altitude is not clear. Presumably the hypoxic stimulus to breathe is one determinant of ventilation at high altitude, but there might be other factors.

One other factor could be the pattern of breathing. Rapid, shallow breathing would be associated with a large dead-space ventilation, which is a component of total ventilation. In the young Sherpa at low altitude (1,377 m) the respiratory frequency was higher, whereas tidal volume was lower than in the trekkers [610 vs. 830 ml ($P < 0.01$)] (14). Sherpas also had lower tidal volume:vital capacity ratios. Studies in Leadville and South America have also reported rapid, shallow breathing and hyperventilation during normoxia (26, 27, 30, 33). The breathing pattern in Sherpas changes with altitude years. As the A value and respiratory frequency declined with increasing altitude years, the tidal volume increased (Fig. 5). A comparison between populations should not be confined to total ventilation, which is influenced by many factors. Measurements are needed that also take into account breathing pattern, alveolar and dead-space ventilation, and metabolic rate.

Another confounding factor that may influence comparison of ventilation between populations is hypoxic ventilatory depression. This phenomenon has been described frequently in high-altitude natives, although the mechanism is unclear. Six of the eight Sherpas to whom we gave 100% oxygen for 1 min at

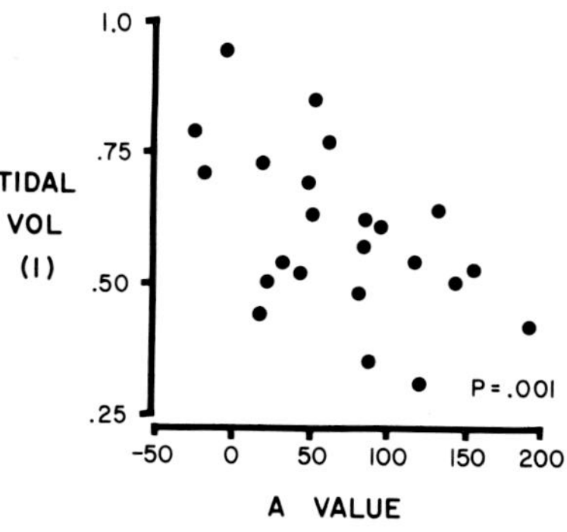

FIG. 5. Relationship in Sherpas of tidal volume to hypoxic ventilatory response measured as parameter *A*. Tidal volume is increased in those with lower *A* values.

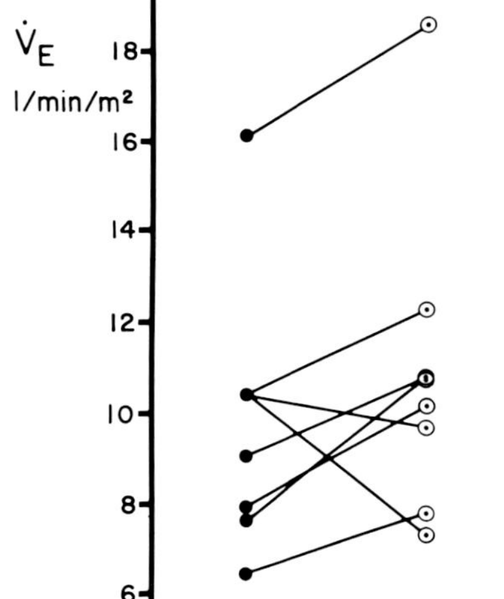

FIG. 6. Minute ventilation ($\dot{V}_E$) during air breathing and 100% O_2 breathing in 8 Sherpas at 4,243 m. Of these, 6 had similar increases in ventilation and 2 had decreased ventilation. [From Hackett et al. (14).]

4,243 m had substantial increases in $\dot{V}_E$ (Fig. 6; 14), similar to a previous report of Sherpas and to reports of Peruvians (35, 56). A paradoxical response in the Sherpa has also been reported during exercise (34). Kryger et al. (26) reported a normal response to oxygen breathing in Leadville natives with blunted drives; they found a paradoxical response only in those who had chronic mountain sickness. Severinghaus et al. (49) have proposed that this paradoxical response is related to blunted chemosensitivity. Thus with long altitude residence there may be both hypoxic depression and blunted hypoxic chemosensitivity. Remarkably the Sherpas we studied maintained their $\dot{V}_E$ values well despite evidence of hypoxic ventilatory depression (14).

Whether Sherpas and Tibetans maintain their pulmonary ventilation during sleep, as in the awake state, is unknown. Sleep apnea has been implicated in the pathogenesis of chronic mountain sickness (25). Weil et al. (58) found that, whereas healthy residents of Leadville have oscillations in Sa_{O_2} during sleep, only those with chronic mountain sickness have prolonged periods of apnea and severe desaturation (Fig. 7). Thus the maintenance of ventilation during sleep may be an important adaptation that helps protect

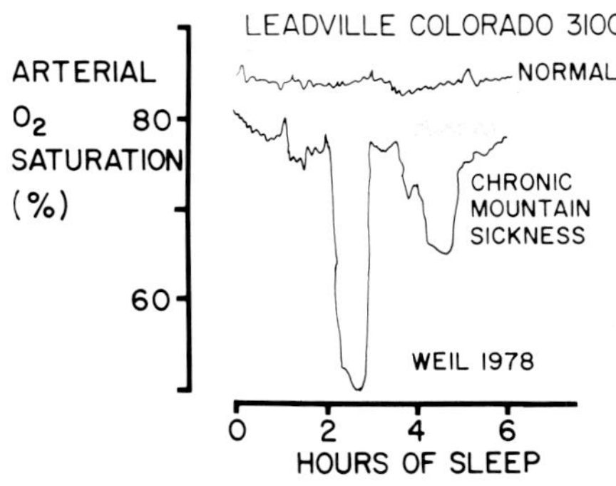

FIG. 7. Arterial O_2 saturation (%) by ear oximeter during sleep in normal residents of Leadville and in those with chronic mountain sickness. Severe and prolonged desaturation occurs in those with chronic mountain sickness. [From Weil et al. (58).]

TABLE 2. *Measurements in Bolivians at La Paz (3,600 m)*

	Hb, g/100 ml blood	$\dot{V}_E$, liters·$min^{-1}·m^{-2}$	f	FVC, liters/min	$P_{A_{O_2}}$, mmHg	$P_{A_{CO_2}}$, mmHg	pHa
Female							
		Natives (22 yr, 159 cm, 55.6 kg, 1.55 m²)					
$n = 10$	15.9 ±0.5	6.7 ±0.4	17.2 ±0.3	2.4 ±0.1	57.1 ±0.8	36.3 ±0.9	7.39 ±0.01
		Newcomers (31 yr, 164 cm, 62.7 kg, 1.67 m²)					
$n = 7$	14.9 ±0.2	6.5 ±0.3	18.1 ±0.5	2.2 ±0.05	61.8 ±1.3	29.1 ±0.7	7.48 ±0.02
Male							
		Natives (25 yr, 167 cm, 61.9 kg, 1.68 m²)					
$n = 14$	16.5 ±0.2	6.0 ±0.3	16.7 ±0.8	2.9 ±0.1	59.9 ±0.5	33.2 ±0.6	7.42 ±0.01
		Newcomers (32 yr, 174 cm, 77.3 kg, 1.91 m²)					
$n = 13$	15.8 ±0.3	6.3 ±0.2	15.1 ±0.7	2.7 ±0.1	60.1 ±0.9	30.7 ±1.2	7.46 ±0.01

Values are means ± SE; barometric pressure was 490 mmHg. Hb, hemoglobin concentration; $\dot{V}_E$, minute ventilation; f, respiratory frequency; FVC, forced vital capacity; $P_{A_{O_2}}$, alveolar partial pressure of O_2; $P_{A_{CO_2}}$, alveolar partial pressure of CO_2; pHa, arterial pH; n, no. of subjects. Anthropometric data reflect age, height, weight, and body surface area, respectively.

against chronic mountain sickness. Whether or not maintaining ventilation during sleep is characteristic of a population well adapted to high altitude deserves study.

Thus far the discussion has not considered women at high altitude. Cudkowicz et al. (10) in Bolivia made an extensive study of women and men in La Paz at 3,600 m and in Chacaltoya at 5,200 m. Selected data from La Paz (Table 2) suggests that the native women have lower hemoglobins, higher vital capacities per unit of body surface area, and higher $P_{A_{CO_2}}$ values than the men. The values of P_{CO_2} in the native women were also higher than in newcomer women who had been in La Paz at least 3 wk. For comparison Table 3 describes women studied on the Tibetan Plateau (59). Ages are not available for the Tibetan study, nor was the duration of residence reported. However, more

TABLE 3. *Hemoglobin concentrations in native highlander workers on Tibetan Plateau (3,800 m) and sojourners at Lhasa (3,658 m)*

	Female	Male
Natives	12.10 ± 0.10 (n = 300)	14.04 ± 0.09 (n = 294)
Sojourners	14.84 ± 0.10 (n = 318)	16.95 ± 0.08 (n = 496)

Values are g/100 ml blood (means ± SE). n, No. of subjects.

TABLE 4. *Incidence of excessive polycythemia in Lhasa workers*

	Female		Male		
	%	Actual no. of cases	%	Actual no. of cases	P
Natives	0	0	1	2	<0.05
Sojourners	16	6	13	90	<0.01

Percent values are no. of cases per 100 population.

Chinese persons were investigated than Bolivians. A comparison suggests that the native Tibetan woman has a lower hemoglobin than the native Bolivian woman at an equivalent altitude of approximately 3,600 m.

In addition, data from the Tibetan population (59) were examined with the value 20 g/100 ml blood or more as an index of excessive polycythemia (Table 4). In both Tibetan natives and lowland Chinese sojourners in Tibet, there were fewer cases of excessive polycythemia among female workers than among male workers. The difference was striking for the sojourners. Sojourners appeared to have more excessive polycythemia than the natives, and the difference was particularly striking among the males. Thus the excessive polycythemia was not observed among native females but was rather frequent among sojourning males. The number of persons in the Tibetan cohort is impressive; the likelihood that ventilatory parameters are important in mediating these differences is discussed in chapter 15.

Ventilatory adaptations among women are especially important during pregnancy. Experience in Leadville indicates that maternal ventilation helps preserve arterial oxygenation during pregnancy (39), and the oxygenation appears to be a determinant of infant birth weight (40). Furthermore low birth weight appears to be a major contributor to infant mortality at high altitude, particularly when combined with prematurity (36). Information from Tuctu, Peru, at 4,400 m in the Andes (16) showed in 13 measurements during gestation in three women that end-tidal P_{CO_2} was 23 ± 0.6 mmHg (means ± SE) and that P_{O_2} was 59 ± 0.8 mmHg. These values suggested hyperventilation compared to 12 measurements in four nonpregnant women (Pa_{CO_2} = 28 ± 0.4; Pa_{O_2} = 51 ± 0.7). Therefore high-altitude pregnancy might be associated with some change that induces hyperventilation. Whether this represents merely the normal hyperventilation of pregnancy or whether some other mechanisms are involved is not known. Furthermore one wonders how pregnancy affects the drives to breathe in high-altitude populations. (Clearly this is a "pregnant" area for further work in high-altitude physiology!)

In assessing the relative importance of pulmonary ventilation in adaptation to altitude, other adaptive phenomena must also be considered. For example, increased capillarization of tissues would decrease the path length for oxygen diffusion to the cell. Increased mitochondrial density (43) could facilitate the intracellular utilization of oxygen. Such adaptations and others could moderate the need for a large increase in ventilation. The relative contribution of these different adaptive processes in Sherpas and other high-altitude natives is largely unexplored. Sherpas, like Tibetans, appear remarkably free from acute mountain sickness when moving from low to high altitudes, although sporadic cases have been reported and those usually at extreme altitudes (6). The 25 Sherpas we studied had a history of 47 expeditions, and they carried heavy loads at extreme altitudes. None had ever had acute mountain sickness. In addition, although Sherpas now move from low altitude to their homeland quickly because of air transportation, no cases of high-altitude pulmonary edema have been reported. Excessive polycythemia, or chronic mountain sickness, has to our knowledge not been reported among Sherpas, but the incidence seems to be relatively high in Leadville. Of a population of approximately 4,000 men, 50 men over 40 yr of age are identified as suffering from chronic mountain sickness (26). Like the Leadville residents, who are relative newcomers to altitude, Chinese and Hispanic peoples who move to high altitudes also have an incidence of chronic mountain sickness that seems to be greater than the true high-altitude natives. In fact several investigations have examined Sherpa blood and reported a remarkable lack of excessive polycythemia without any difference in P_{50} or hemoglobin (4, 38, 47). From what is known of chronic mountain sickness, a maintained pulmonary ventilation would seem to be protective, as it is in acute mountain sickness. More data are needed to document that the Sherpa and the Tibetan are comparatively immune from acute mountain sickness and excessive polycythemia. If such immunity is found we need to establish the relative roles of ventilatory and tissue adaptations.

REFERENCES

1. BEALL, C. M. A comparison of chest morphology in high altitude Asian and Andean populations. *Hum. Biol.* 54: 145–163, 1982.
2. BYRNE-QUINN, E., I. E. SODAL, AND J. V. WEIL. Hypoxic and hypercapnic ventilatory drives in children native to high altitude. *J. Appl. Physiol.* 32: 44–46, 1972.
3. CERRETELLI, P. Limiting factors to oxygen transport on Mount Everest. *J. Appl. Physiol.* 40: 658–667, 1976.
4. CERRETELLI, P. Gas exchange at high altitude. In: *Pulmonary Gas Exchange*, edited by J. B. West. New York: Academic, 1980, vol. 2, p. 106.
5. CHIODI, H. Respiratory adaptations to chronic high altitude hypoxia. *J. Appl. Physiol.* 10: 81–87, 1957.
6. CLARKE, C., AND J. DUFF. Mountain sickness, retinal hemorrhage, and acclimatization on Mount Everest in 1975. *Br. Med. J.* 2: 495–497, 1976.
7. COLLINS, D. D., C. H. SCOGGIN, C. W. ZWILLICH, AND J. V. WEIL. Hereditary aspects of decreased hypoxic response. *J. Clin. Invest.* 62: 105–110, 1978.
8. CRUZ, J. Mechanisms of breathing in high altitude and sea level subjects. *Respir. Physiol.* 17: 146–161, 1973.
9. CRUZ, J. C., AND R. J. ZEBALLOS. Influencia racial sobre la respuesta ventilatoria a la hipoxia e hipercapnia. *Acta Physiol. Lat. Am.* 25: 23–32, 1975.
10. CUDKOWICZ, L., H. SPIELVOGEL, AND G. ZUBIETA. Respiratory studies in women at high altitude (3600 M or 12,200 ft and 5200 M or 17,200 ft). *Respiration* 29: 393–426, 1972.
11. FORSTER, H. V., J. A. DEMPSEY, M. L. BIRNBAUM, W. G. REDDAN, J. THODEN, R. F. GROVER, AND J. RANKIN. Comparison of ventilatory responses to hypoxic and hypercapnic stimuli in altitude-sojourning lowlanders, lowlanders residing at altitude and native altitude residents. *Federation Proc.* 28: 1274–1279, 1969.
12. FRISANCHO, R. A., AND P. T. BAKER. Altitude and growth: a study of the patterns of physical growth of a high altitude Peruvian Quechua population. *Am. J. Phys. Anthropol.* 32: 279–292, 1970.
13. GROVER, R. F., J. T. REEVES, E. B. GROVER, AND J. E. LEATHERS. Muscular exercise in young men native to 3,100 m altitude. *J. Appl. Physiol.* 22: 555–564, 1967.
14. HACKETT, P. H., J. T. REEVES, C. D. REEVES, R. F.

GROVER, AND D. RENNIE. Control of breathing in Sherpas at low and high altitude. *J. Appl. Physiol.: Respirat. Environ. Exercise Physiol.* 49: 374–379, 1980.

15. HACKETT, P. H., D. RENNIE, S. E. HOFMEISTER, R. F. GROVER, AND J. T. REEVES. Fluid retention and relative hypoventilation in acute mountain sickness. *Respiration* 43: 321–329, 1982.
16. HELLEGERS, A., J. METCALF, W. E. HUCKABEE, H. PRYSTOWSKY, G. MESCHIA, AND D. H. BARRON. Alveolar P_{CO_2} and P_{O_2} in pregnant and nonpregnant women at high altitude. *Am. J. Obstet. Gynecol.* 82: 241–245, 1961.
17. HIRSHMAN, C. A., R. E. McCULLOUGH, AND J. V. WEIL. Normal values for hypoxic and hypercapnic ventilatory drives in man. *J. Appl. Physiol.* 38: 1095–1098, 1975.
18. HU, S.-T., S.-Y. HUANG, S.-C. CHU, AND C.-F. PA. Peripheral chemoresponsiveness at sea level and susceptibility of acute mountain sickness. In: *Geological and Ecological Studies of Qinghai-Xizang Plateau.* New York: Gordon & Breach, p. 1357–1362, 1981. [Proc. Symp. Qinghai-Xizang (Tibet) Plateau, Beijing, China, vol. 2.]
19. HUANG, S.-Y., Z.-Z. GU, C.-F. PA, AND S.-T. HU. Ventilatory control in Tibetan highlanders. In: *Geological and Ecological Studies of Qinghai-Xizang Plateau.* New York: Gordon & Breach, p. 1363–1369, 1981. [Proc. Symp. Qinghai-Xizang (Tibet) Plateau, Beijing, China, vol. 2.]
20. HURTADO, A. Animals in high altitudes: resident man. In: *Handbook of Physiology. Adaptation to the Environment,* edited by D. B. Dill and E. F. Adolf. Washington, DC: Am. Physiol. Soc., 1964, sect. 4, chapt. 54, p. 843–860.
21. HURTADO, A., AND H. ASTE-SALAZAR. Arterial blood gases and acid-base balance at sea level and at high altitudes. *J. Appl. Physiol.* 1: 304–325, 1948.
22. KENNTNER, G. Gebrauche und leistungsfahigkeit im tragen von lasten bei bewohnern des sudlichen Himalaya. *Z. Morphol. Anthropol.* 61: 125–169, 1969.
23. KING, A. B., AND S. M. ROBINSON. Ventilation response to hypoxia and acute mountain sickness. *Aerosp. Med.* 43: 419–421, 1972.
24. KRONENBERG, R. S., AND C. W. DROGE. Ventilatory and heart rate responses to hypoxia and hypercapnia with aging in normal man. *J. Clin. Invest.* 52: 1812–1819, 1973.
25. KRYGER, M., R. GLAS, D. JACKSON, R. E. McCULLOUGH, C. SCOGGIN, R. F. GROVER, AND J. V. WEIL. Impaired oxygenation during sleep in excessive polycythemia of high altitude: improvement with respiratory stimulation. *Sleep* 1: 3–17, 1978.
26. KRYGER, M., R. McCULLOUGH, R. DOEKEL, D. COLLINS, J. V. WEIL, AND R. F. GROVER. Excessive polycythemia of high altitude: role of ventilatory drive and lung disease. *Am. Rev. Respir. Dis.* 118: 659–665, 1978.
27. LAHIRI, S. Alveolar gas pressures in man with life-time hypoxia. *Respir. Physiol.* 4: 373–386, 1968.
28. LAHIRI, S., J. S. BRODY, E. K. MOTOYAMA, AND T. M. VELASQUEZ. Regulation of breathing in newborns at high altitude. *J. Appl. Physiol.: Respirat. Environ. Exercise Physiol.* 44: 673–678, 1978.
29. LAHIRI, S., R. G. DELANEY, J. S. BRODY, M. SIMPSEN, T. VELASQUEZ, E. K. MOTOYAMA, AND C. POLGAR. Relative role of environmental and genetic factors in respiratory adaptation to high altitude. *Nature London* 261: 133–135, 1976.
30. LAHIRI, S., AND N. H. EDELMAN. Peripheral chemoreflexes in the regulation of breathing of high altitude natives. *Respir. Physiol.* 6: 375–385, 1969.
31. LAHIRI, S., F. F. KAO, T. VELASQUEZ, C. MARTINEZ, AND W. PEZZIA. Respiration of man during exercise at high altitude: highlanders vs. lowlanders. *Respir. Physiol.* 8: 361–375, 1970.
32. LAHIRI, S., AND J. S. MILLEDGE. Acid-base in Sherpa altitude residents and lowlanders at 4880 M. *Respir. Physiol.* 2: 323–324, 1967.
33. LAHIRI, S., J. S. MILLEDGE, H. P. CHATTOPADHYAY, A. K. BHATTACHARYYA, AND A. K. SINHA. Respiration and heart rate of Sherpa highlanders during exercise. *J. Appl. Physiol.* 23: 545–554, 1967.
34. LAHIRI, S., J. S. MILLEDGE, AND S. C. SØRENSEN. Ventilation in man during exercise at high altitude. *J. Appl. Physiol.* 32: 766–769, 1972.
35. LEFRANCOIS, R., H. GAUTIER, AND P. PASQUIS. Ventilatory oxygen drive in acute and chronic hypoxia. *Respir. Physiol.* 4: 217–228, 1968.
36. McCULLOUGH, R. E., J. T. REEVES, AND R. L. LILGEGREN. Fetal growth retardation and increased infant mortality at high altitude. *Arch. Environ. Health* 32: 36–39, 1977.
37. MILLEDGE, J. S., AND S. LAHIRI. Respiratory control in lowlanders and Sherpa highlanders at altitude. *Respir. Physiol.* 2: 310–322, 1967.
38. MONPURGO, G., P. ARESE, A. BOSIA, G. P. PESEARMONA, M. LUZZANA, G. MODIANO, AND S. K. RONJIT. Sherpas living permenantly at high altitude: a new pattern of adaptation. *Proc. Natl. Acad. Sci. USA* 73: 747–751, 1976.
39. MOORE, L. G., D. JAHNIGEN, S. S. ROUNDS, J. T. REEVES, AND R. F. GROVER. Maternal hyperventilation helps preserve arterial oxygenation during high-altitude pregnancy. *J. Appl. Physiol.: Respirat. Environ. Exercise Physiol.* 52: 690–694, 1982.
40. MOORE, L. G., S. S. ROUNDS, D. JAHNIGEN, R. F. GROVER, AND J. T. REEVES. Infant birth weight is related to maternal arterial oxygenation at high altitude. *J. Appl. Physiol.: Respirat. Environ. Exercise Physiol.* 52: 695–699, 1982.
41. MUELLER, W. H., F. MURILLO, H. PALAMINO, M. BADZIOCH, R. CHAKRABORTY, P. FUERST, AND W. J. SCHULL. The Aymara of Western Bolivia. V. Growth and development in an hypoxic environment. *Hum. Biol.* 52: 529–546, 1980.
42. MUELLER, W. H., F. YEN, F. ROTHLAMMER, AND W. J. SCHULL. A multinational Andean genetic and health program. VI. Physiological measurements of lung function in an hypoxic environment. *Hum. Biol.* 50: 489–513, 1978.
43. OU, L. C., AND S. M. TENNEY. Properties of mitochondria from hearts of cattle acclimatized to high altitude. *Respir. Physiol.* 8: 151–165, 1970.
44. PAWSON, I. G. Growth characteristics of population of Tibetan origin in Nepal. *Am. J. Phys. Anthropol.* 47: 473–482, 1977.
45. PUGH, L. G. C. E., M. B. GILL, S. LAHIRI, J. S. MILLEDGE, M. P. WARD, AND J. B. WEST. Muscular exercise at great altitudes. *J. Appl. Physiol.* 19: 431–440, 1964.
46. REEVES, J. T., J. K. ALEXANDER, R. F. GROVER, S. Y. HUANG, J. T. MAHER, R. E. McCULLOUGH, L. G. MOORE, J. B. SAMPSON, AND J. V. WEIL. Ventilation at rest and exercise at high altitude can be predicted from hypoxic ventilatory drives measured at low altitude (abstr.). *Clin. Res.* 31: 74A, 1983.
47. SAMAJA, M., A. VEICSTEINAS, AND P. CERRETELLI. Oxygen affinity of blood in altitude Sherpas. *J. Appl. Physiol.: Respirat. Environ. Exercise Physiol.* 47: 337–341, 1979.
48. SCOGGIN, C. H., R. D. DOEKEL, M. H. KRYGER, C. W. ZWILLICH, AND J. V. WEIL. Familial aspects of decreased hypoxic drive in endurance athletes. *J. Appl.*

Physiol.: Respirat. Environ. Exercise Physiol. 44: 464–468, 1978.

49. SEVERINGHAUS, J. W., C. R. BAINTON, A. CARCELÈN. Respiratory insensitivity to hypoxia in chronically hypoxic man. *Respir. Physiol.* 1: 308–334, 1966.
50. SIME, F., D. PENALOZA, AND L. RUIZ. Bradycardia, increased cardiac output, and reversal of pulmonary hypertension in altitude natives living at sea level. *Br. Heart J.* 33: 647–657, 1971.
51. SLOAN, A. W., AND M. MASALI. Anthropometry of Sherpa men. *Ann. Hum. Biol.* 5: 453–458, 1978.
52. SØRENSEN, S. C., AND J. W. SEVERINGHAUS. Respiratory sensitivity to acute hypoxia in man born at sea level living at high altitude. *J. Appl. Physiol.* 25: 211–216, 1968.
53. SØRENSEN, S. C., AND J. W. SEVERINGHAUS. Irreversible respiratory insensitivity to acute hypoxia in man born at high altitude. *J. Appl. Physiol.* 25: 217–220, 1968.
54. TENNEY, S. M., AND L. C. OU. Hypoxic ventilatory response to cats at high altitude: an interpretation of "blunting." *Respir. Physiol.* 30: 185–199, 1977.
55. VELASQUEZ, T. Pulmonary function and oxygen transport. In: *Man in the Andes, A Multidisciplinary Study of High Altitude Quechua*, edited by P. Baker and F. Little. Stroudsburg: Dowden, Hutchinson, & Ross, 1976, p. 237–260.
56. VELASQUEZ, T., C. MARTINEZ, W. PEZZIA, AND N. GALLARDO. Ventilatory effects of oxygen in high altitude natives. *Respir. Physiol.* 5: 211–220, 1968.
57. WEIL, J., E. BYRNE-QUINN, I. E. SODAL, G. F. FILLEY, AND R. F. GROVER. Acquired attenuation of chemoreceptor function in chronically hypoxic man at high altitude. *J. Clin. Invest.* 50: 186–195, 1971.
58. WEIL, J. V., M. H. KRYGER, AND C. H. SCOGGIN. Sleep and breathing at high altitude. In: *Sleep Apnea Syndromes*, edited by C. Guilleminault and W. C. Dement. New York: Liss, 1978, p. 119–136.
59. XIE, C.-F., AND S.-X. PEI. Some physiological data on sojourners and native highlanders at 3 different altitudes in Xizang. In: *Geological and Ecological Studies of Qinghai-Xizang Plateau.* New York: Gordon & Breach, p. 1449–1452, 1981. [Proc. Symp. Qinghai-Xizang (Tibet) Plateau, Beijing, China, vol. 2.]

INDEX